Preface

Since the last edition of Dr. Maurer's book the advances in the field of electrophoresis have progressed at an unprecedented pace. The rapid expansion of the technique of isoelectric focusing in both polyacrylamide gel and in charge free agarose has led to a disproportionate increase in this methodology as compared to PAGE.

Similarly, two other areas of electrophoresis have come into their own since in the last thirteen years: that of separation in the presence of detergents for molecular weight studies and two dimensional methods utilizing dissociating methods in the first dimension by isoelectric focusing and denaturing methods in the second dimension with SDS-PAGE.

This shift in direction and emphasis has led to a number of important new applications of polyacrylamide and agarose electrophoresis. However, the format of this edition will parallel that of the previous edition with an emphasis on the practical applications of techniques. It will be apparent that this edition is slanted toward those areas of experience of the authors.

Since the first German edition appeared in early 1968, an average of 3000 papers a year have appeared in this field. It is both impossible and unnecessary to cite them all. Obviously, many no longer warrant citation, since they are outmoded, or incorrect in the light of recent work. However, we have attempted to sift through, at least, most of that vast amount of literature pertinent to the aims of the book, with the realization that much may have been overlooked and that some even may be outdated by the time of publication.

To provide the potential for rapid updating of this book it was written and formatted on a personal computer using the Gutenberg™ program on an Apple II™ computer in an attempt to keep the cost at a reasonable level. In fact the day is fast approaching where future books well may be obtained directly from the publisher on disk. While this might seem to some a bit futuristic, this book was written between Boston and Charleston in just such a manner.

Acknowledgements

The authors would like to take this opportunity to express their deep appreciation to our many collegues who were so generous with illustrations and information that made this effort possible. We would like to acknowledge specifically Drs. Andreas Chrambach, Berthold Radola, Carl Merril, Klaus Altland, Michael Dunn, Pier Georgio Righetti, Joachim Klose, John Fawcett and Leigh Anderson for their kind considerations, the time spent on many phone calls and letters that provided much of the stimulation and material for this book.

The senior authors would also take this opportunity to recognize their wives, Carol and Judy who have become literally computer widows for the last year while we have been writing this book. Our many thanks and deep appreciation for their fore-bearance and understanding, when for hours our only utterances were expletives when we blew a disk.

We also would like to express our appreciation to Walter de Gruyter and specifically to Dr. Rudolf Weber for his support of this concept of producing a book aimed at reducing costs and decreasing production time and to Micro-Map Inc. for allowing the use of major parts of Chapters 3 and 5.

Finally, we would like to express our admiration for those unsung heroes, the type-setters, whose trials and tribulations only become apparent when one attempts to format a book himself. Also a word of appreciation is in order to Mr. Johann Wagner, the developer of the Gutenberg™ program from Gutenberg Software Ltd., Scarborough, Ontario, Canada, which allowed amateurs to attempt such a task.

Isle of Palms, South Carolina, April 1984 — R. C. A.

Boston, Massachusetts, April 1984 — C. A. S.

Gel Electrophoresis and Isoelectric Focusing of Proteins

Selected Techniques

R. C. Allen · C. A. Saravis · H. R. Maurer

Gel Electrophoresis and Isoelectric Focusing of Proteins

Selected Techniques

Walter de Gruyter
Berlin · New York 1984

Robert C. Allen, Ph. D.
Professor and Chairman, Department of Laboratory Animal Medicine
Professor of Pathology at the Medical University of South Carolina
171 Ashley Avenue
Charleston, SC 29425, USA

Calvin A. Saravis, Ph. D.
Assistant Director of the GI Laboratory of the Mallory Institute
Principle Associate in Surgery Harvard University, Cambridge, MA
Associate Professor of Pathology at Boston University
784 Massachusetts Avenue
Boston, MA 02118

Professor Dr. H. Rainer Maurer
Pharmazeutisches Institut der FU Berlin
Königin-Luise-Straße 2 – 4
D-1000 Berlin 33
West Germany

CIP-Kurztitelafnahme der Deutschen Bibliothek

Allen, Robert C.:
Gel electrophoresis and isoelectric focusing of proteins:
selected techniques / R. C. Allen ; C. A. Saravis ; H. R. Maurer. –
Berlin ; New York : de Gruyter, 1984.
ISBN 3-11-007853-8
NE: Saravis, Calvin A.: ; Maurer, Hans Rainer

Library of Congress Cataloging in Publication Data

Allen, R. C. (Robert Carter), 1930 –
Gel electrophoresis and isoelectric focusing of proteins.
Bibliography: p.
Includes index.
1. Proteins--Analysis. 2. Electrophoresis, Polyacrylamide gel. 3. Isoelectric focusing.
I. Saravis, C. A. (Calvin Albert), 1930 – . II. Maurer, H. Rainer. III. Title.
QP551.A435 1984 574.19'245 84-12694
ISBN 3-11-007853-8

ISBN 3-11-007853-8 Verlag Walter de Gruyter Berlin · New York
ISBN 0-899-25002-5 Walter de Gruyter, Inc. New York · Berlin

Printing: Druckerei Gerike GmbH, Berlin. Binding: Dieter Mikolai, Berlin.

Frontispiece

The quest for ever increased improvement in resolution in electrophoretic separations, perhaps pervades too many of our waking hours and bench time. It is refreshing that nature has already pointed the way in her iniminitable fashion, with the pseudo-silver stain patterns developed millenia earlier and shown in the accompanying illustration.

The brown banded pattern found on the jackknife clam *(Ensis minor),* indigenous to the southern coast of the USA. These mimic silver stained gels, or more properly *vice versa.*

Contents

1. GEL SUPPORT MEDIA

1.1. Polyacrylamide Gel

1.1.1. Formation and structure of the gel

Polyacrylamide gel is the polymerization product of the monomer acrylamide, CH_2=CH-CO-NH_2 and a cross linking comonomer, most commonly, N,N'-methylene-bis-acrylamide (Bis), CH_2=CH-CO-NH-CH_2-NH-CO-CH=CH_2 (1-3). The three dimensional network is formed by cross-linking of polyacrylamide chains growing side by side by the mechanism of vinyl polymerization. This leads to the development of random polymer coils. The advantages over previous support media include that polyacrylamide produces a clear gel and that the pore size may be varied over a wide range simply by increasing the monomer concentration, thus providing more selective sieving of macromolecules.

However, it must be remembered that this material was originally developed and used for fibers and plastics, not for biological separations. It is therefore critical to understand some of the properties of the monomer, prior to discussing its various uses, advantages and disadvantages as a sieving support media for electrophoretic separations.

1.1.2. Acrylamide monomer: Practical considerations

Acrylamide monomer is produced by the reduction of acrylonitrile by either liquid ammonia or by calcium bisufite. The former method is used in the USA by American Cyanamide and the latter in Europe by the Shell company. Both materials are suitable for electrophoretic and isoelectric focusing separations after further purification by recrystallization. Little has appeared in the literature, other than a summary of extensive discussions at the 1972 Tübingen meeting (4), concerning the pitfalls of using non-purified, or purified acrylamide that has been stored too long, as a support medium for electrophoretic separation and many may not have encountered these reports.

1.1.2.1. Purification of the monomer

Acrylamide monomer may be purified effectively by a single step chloroform recrystallization (5). The procedure is as follows:

Table 1. Monomer recrystallization procedure

1.	Add 45 g of monomer to 90 ml of chloroform heated to 60 °C in hot water bath. Dissolve all acrylamide and hold at this temperature for 5 min. Swirl to assure complete solution of all soluble monomer.
2.	Filter through Whatman #1 filter paper, where the funnel is placed in a heating jacket holding the solution being filtered at 60 °C, otherwise the monomer will crystallize out on the filter paper and in the funnel. The filtrate should be collected in a heating jacketed Florence flask maintained at 60 °C.
3.	After filtration, cool the Florence flask containing the filtrate in an ice bath to no lower than 22 °C and immediately filter under *vacuo* through Whatman #1 filter paper in a Buchner funnel until all liquid is removed from the crystals.
4.	Remove the wet crystals fron the Buchner funnel with a spatula and spread on a large sheet of filter paper to dry. Occassionaly respread the crystals and crush out any lumps. The yield should be about 60 per cent. Discard the chloroform-monomer residual as appropriate for contaminated waste.
5.	When there is no residual odor of chloroform, place the dried crystals in a brown glass or plastic container and store at -20 °C under dessication. New monomer should be prepared a least weekly.

Note: All procedures should be carried out in a fume hood, since acrylamide monomer is considered to be an accumulative neurotoxin.

In the recrystallization of the monomer the temperature should not be dropped lower than 22 °C, otherwise additional unknown material will co-precipitate giving an undesirable product. For example, when the crystallization is carried out a 0 °C and gels compared on the same sample from both preparations, it was found that prealbumin esterases would resolve in the former, but not in the latter (4).

Recrystallized acrylamide is more self-reactive than the less pure electrophoresis grade. It breaks down into a series of compounds including the following: free acrylic acid, polyacrylic acid, poly-acrylamide, ammomia, and ß',ß'',ß''' Nitrilotrispropionamide, a strong base. Thus, prolonged storage of the recrystallized monomer should be avoided. Commercially purified monomer available from Bio-Rad, Polysciences, Serva, and others is a suitable alternative to recrystallizing electrophoresis grade monomer, such as produced by Eastman Kodak. However, these too will begin to break down with time. Normally in our experience the monomer is adequate for a least one month for more sensitive techniques such as high voltage gradient focusing on ultrathin-layer gels followed by diammine silver staining, if kept frozen under dessication after first being opened. Bisacrylamide can be recrystallized from acetone; however, the small amounts normally used in relation to the monomer, do not seem to require this step if fresh commercial BIS is used.

1.2. Relationship of Monomer and Crosslinker to Pore Size

In attempts to explain the retardation of a particle by a gel network, a series of models have been proposed to describe pores in polyacrylamide gels, from the observations of Raymond and Nakamichi in 1962 (6), to the present. This is obviously a critical point in any separation procedure in which one wishes to utilize the size of a molecule as a parameter in the separation process.

The effective pore size of a polyacrylamide gel is an inverse function of "total monomer concentration", (% T), defined as the sum of the concentrations of acrylamide monomer and the crosslinking agent.

A polyacrylamide gel of low concentration can be thought of as being a pore-meshed , three dimensional lattice of long polyacrylamide chains tied at distant intervals by "knots" of crosslinker. A number of authors have described the geometric configuration of this network and the more recent studies of Ziabicki (7) indicate the formation of unentangled and entangled chains as shown below in Fig. 1.

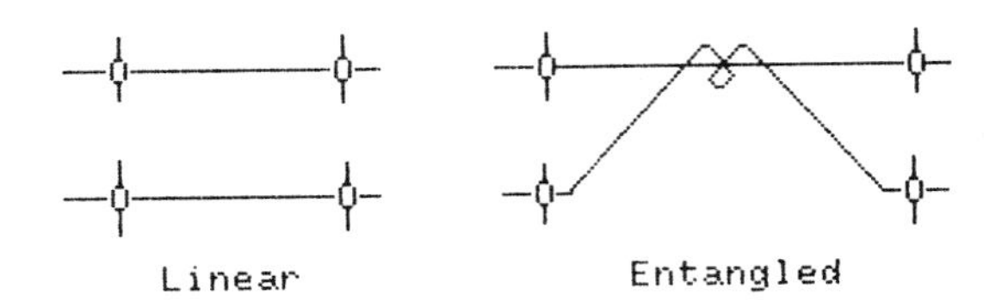

Fig. 1. Formation of entangled chains in the polymerization of polyacrylamide gels (from Ziabicki) (7).

Righetti (8) has proposed a somewhat different structure from the un-crosslinked to a pure crosslinked gel, which is depicted in Fig. 2. below. .

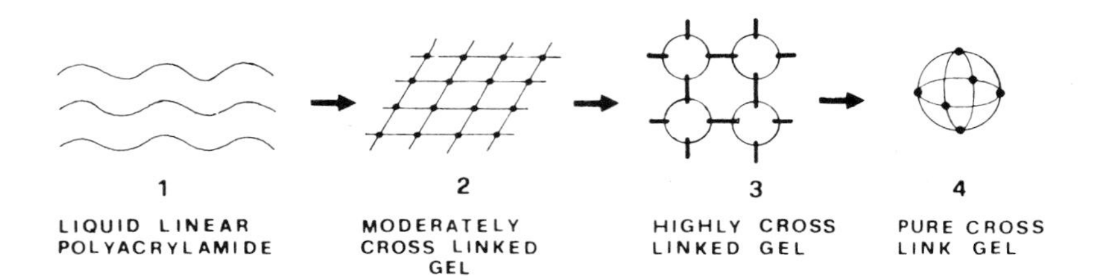

Fig. 2. Proposed variations of gel structure from unpolymerized monomer to fully cross linked gel. As the crosslinking increases there is a shortening and thickening of the gel fibers with the growth of beads (from Righetti) (8).

When per cent T is increased and a low per cent C maintained, the frequency of chains increases, thereby reducing protein mobility due to frictional retardation. This has been showm for gel concentrations up to 40 and even 50 per cent T at 5 per cent C (9,10). As cross linking is increased, a regular array of chains, tied by "knots" is formed which, at five per cent C, has maximal sieving properties, *i. e.* a minimum pore size with a constant value of T. As per cent C is increased above 10 per cent there is a diminishing number of singlets, which favors irregularities in the gel resulting from more doublet and loop formation. As these increase, two simultaneous events occur: the linear chains grow shorter and thicker and the "knots" grow larger. This produces larger pore sizes as depicted in Fig. 2. and shown in Fig. 5. When the system approaches a composition of pure crosslinkage, the chains disappear and the resulting loops grow into beads or spheres. This explains the high porosity of extremely high per cent C gels, which at 50 per cent C effectively resemble gel filtration media (11). This may be predicted also from the development of equations linking pore size p to gel concentration. These equations are shown in Fig. 3. below.

$$\bar{p} = Kd / \sqrt{C} \quad (6)$$

$$\bar{p} = \frac{K'd}{C} + K'' \quad (12)$$

$$\bar{p} \propto 1 / \sqrt[3]{C} \quad (10)$$

$$\bar{p} = 140.7 \times C^{-0.7} \quad (13)$$

Fig. 3. Proposed equation linking the mean pore size to the gel concentration; Raymond and Nakimichi (6), Toombs (12), Rodbard and Chrambach (10) and Righetti *et al.* (13). The last equation is for agarose matrices.

Campbell *et al.* (14) have reported the use of a high concentration T (40 %) combined with a high level of crosslinking (12.5 %) which yields clear gels capable of restricting the passage of small proteins. These were in the form of gradient gels in which T was raised from 3 to 40 per cent and C from 4 to 12.5 per cent. In such gels both at acid and basic pHs all proteins were retained in the gels after even prolonged electrophoresis and the zones were found to be sharper than with conventional crosslinking percentages of BIS.

Rüchel and Brager (15) and Rüchel *et al.* (16) have eloquently demonstrated the effects of the interaction between the concentations of per cent T and C when BIS is used as the crosslinker by freeze-etched polyacrylamide gels subjected to transmission-electron microscopy, as shown in the following two figures:

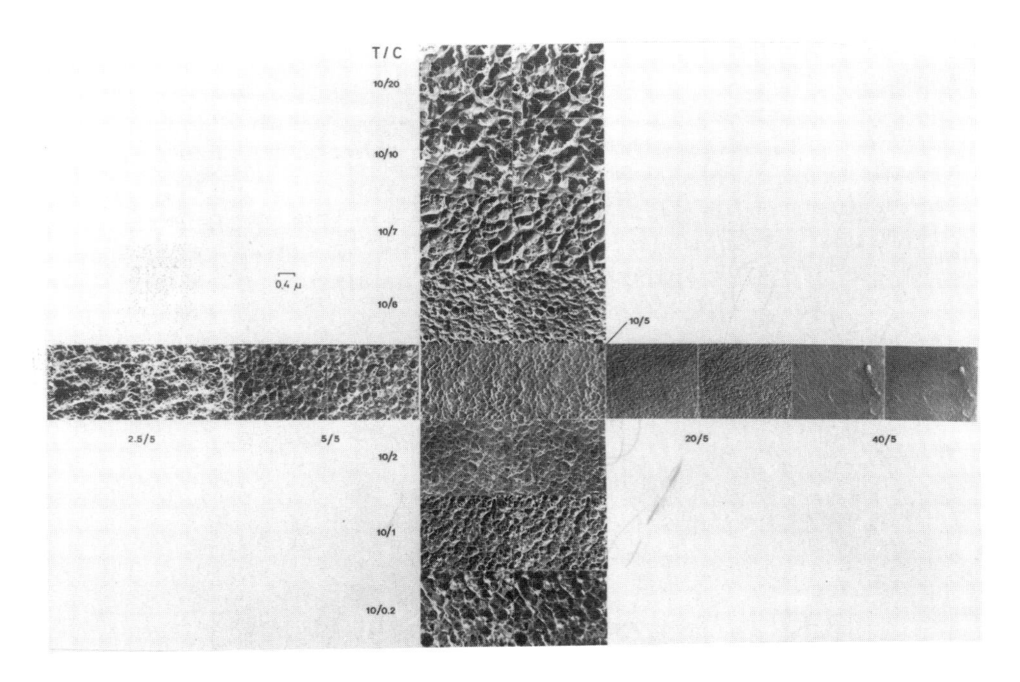

Fig. 4. TEM images of polyacrylamide gels with varying concentrations of total acrylamide (T) and varying concentrations of crosslinker BIS (C). The values shown are T/C; the bar equals 400 nm. In the horizontal series, C remains constant and T increases from left to right. In the vertical series, T remains constant and C is increased from the bottom to the top. (from Rüchel *et al.*) (16)

The series shown above indicates an almost parabolic function of gel structure with a minimum pore size at approximately 5 per cent C. This confirms previous findings concerning the relationship between crosslinkage and sieving in polyacrlamide gels.

Pore size reproducibility depends on the accuracy and purity of monomer stock solutions and of their pipetting. In practice one can monitor the accuracy of stock solutions by spectrophotometric analysis in the range of 230 - 260 nm (17). Even with properly purified reagents and accurate monomer concentrations, the polymer formed will not be reproducible unless the conversion of monomer to polymer is nearly quantitative; *i.e.* at least 95 per cent and preferably 99 per cent conversion should be achieved.

An additional important observation of Rüchel *et al.* (16) was that polymerization at an interface of the gel and the casting chamber surface produces much larger pore sizes in this region as shown in Fig. 5. below.

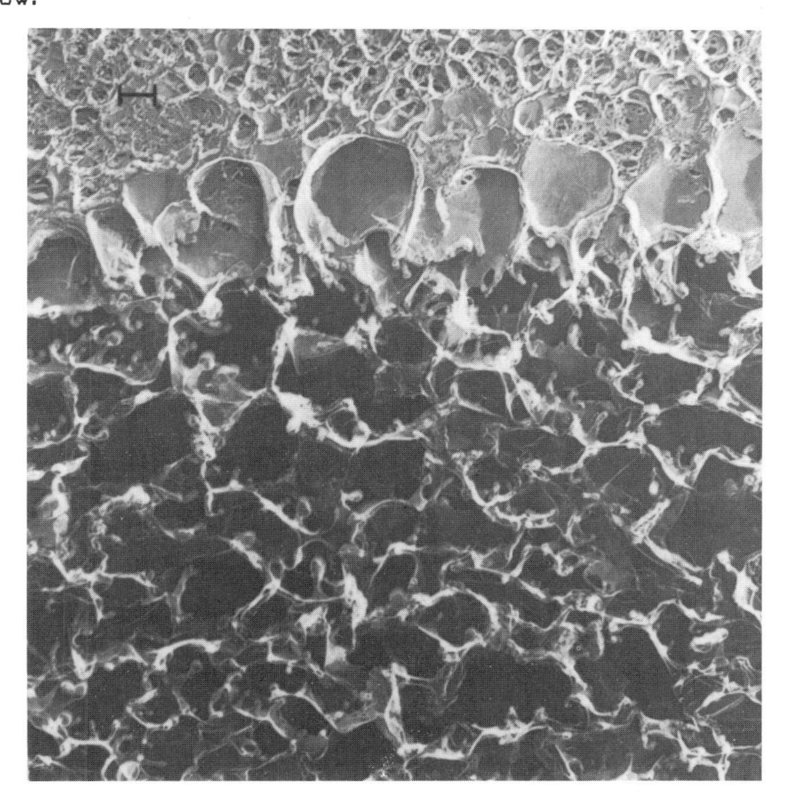

Fig. 5. TEM image of a 10 per cent T, 1 per cent C polyacrylamide gel; the bar equals 100nm. The bottom of the gel is that adjacent to the glass surface of a chamber. The surface at the interface shows a much larger pore size than deeper into the gel.(from Rüchel *et al.*) (16).

The consequence of this latter finding probably has little effect on relatively thick rod or slab gels. However, in ultrathin-layer gels of 100μ or less , this may provide an effective opening of the pore size, quite desirable in polyacrlamide gel isoelectric focusing (PAGIF). It may also lead to problems insurfacesmearing of lipoproteins 100μ gels where lipoprotein - esterase complexes tend to run on the surface, while they do not in thicker 250 μ PAGIF gels (18).

1.2.1. Polymerization procedure cautions

a. Initiators of polymerization

The initiators presently used in polyacrylamide gel electrophoresis include ammonium persulfate (AP), potassium persulfate (KP) and N,N,N^1,N^1 -tetramethylethylenediamine (TEMED). Also riboflavin (RN) is commonly used for photo-activated polymerization in the ultra violet region. However, caution must be exercised when this is the sole initiator used for polymerization. Righetti *et al.* (19) have shown that photopolymerization with RN requires at least 8 hours to reach 95 per cent conversion of the monomer to the polymer. Previous reports of one hour (17,20,21) produce only 50 to 60 per cent conversion of monomer to polymer. This not only leads to irreproducible gels, but also the free monomer can react, at the appropriate pH with amino acid residues in proteins, such as His, Lys, Tyr, Cys and the free amino acid terminus (19).

It appears, from studies with a single per cent T and C and pH that the degree of conversion from monomer to polymer is proportional to the concentrations of AP, RN and TEMED each raised to an exponent near unity (22). To obtain 99 per cent conversion of monomer to polymer at a minimal catalyst concentration, these authors suggest a combination of all three initiators. It is also advantageous then to to use a higher concentration of that catalyst that migrates in the same direction, but faster than the macromolecule or macromolecules of interest. Toward this aim, an increase in persulfate is reasonable with the elimination of the RN. As a rule of thumb, **10mM** concentrations appear to achieve this goal, with the TEMED being increased in those systems that are more acid; particularly, in isoelectric focusing gels employing acid range pH gradients. These considerations are, of course of greater consequence where precise control of pore size is required such as in SDS-PAGE. Where a larger pore size is desired so that migration will not be inhibited, such as in isoelectric focusing, these considerations are not so important. However, in isoelectric focusing, a prefocusing step is desirable to remove any unpolymerized monomer resulting from less stringent polymerization conditions. It is also important to remove any breakdown products from persulfate, namely sulfate ion, from systems to be run with a continuous buffer. The possibility of an inadvertant moving boundary system occuring, if this is not done, is discussed in Chapter 2 at greater length.

b. Inhibitors

Any compound that may act as a free radical trap will act as a polymerization inhibitor.Thus, to obtain reproducible polymerization rates and gels it is necessary to control the partial pressure of oxygen in the the polymerization mixture. This may be accomplished by timed deaeration under *vacuo*, or by argon gassing. A simple, but effective method is to draw the polymerization mixture up into a 50 ml plastic syringe with a rubber tipped plunger. Seal the open end with one finger and pull back on the plunger as hard as possible and then tap the syringe barrel on a lab bench top edge 8 to 10 times, or until no more air bubbles rise out of the solution. Interestingly, some oxygen must remain in the solution for polymerization to occur and one of the products of polymerization, using either riboflavin or persulfate, is peroxide (4).

Inhibition of polymerization may also occur from the materials used for the polymerization chamber. Pyrex, quartz glass, teflon, polyvinyl chloride and Parafilm have no inhibitory affect. On the other hand, rubber and untreated plastics, such as lucite and hydrophobic polyester films inhibit polymerization at the gel - material surface interface causing lack of adhesion. This can result in potential buffer leakage though the cell in vertical, closed chamber, systems and incomplete polymerization on the exposed surface of horizontal PAGIF gels.

c. Rate of polymerization

All of the above parameters affect the rate of polymerization. Therefore, their adequate control is necessary to produce a reproducible polymer. A polymerization time of 10 min. is suggested by Chrambach and Rodbard (22) for PAGE gel rods. Maurer and Allen (23), have suggested 20 min. for PAGE gel slabs, at which time a strong refractile line should be visible at the gel - water layer interface.

For polyacrylamide isoelectric focusing gels (PAGIF) longer polymerization times are usually required, particularly with ultrathin-layer gels discussed later under the section on gel casting (Chapter 3, section 3.12.3.). In both PAGE gel configurations polymerization will proceed to 95 - 99 per cent completion even at 0°C. Since the polymerization procedure is an exothermic one, care must be taken to control the temperature during polymerization, or risk that the gel swells at the ends, or not adhere properly to the wall of the cell or tube. Flat slabs, with their greater surface for cooling are easier to control in this respect. In the preparation of gradient gels it is important to control the polymerization rate to assure polymerization from the top downward or heat produced convective mixing can occur effectively destroying the pore gradient.

1.2.2. Other crosslinking agents

Since mechanical stability decreases as crosslinking with bisacrylamide is increased, it is advantageous in practice (where separations will be carried out at 0 - 4 °C) to replace BIS by other crosslinking agents, particularly N,N^1,-diallyltartardiamide (DATD) (24).

The effective pore size of DATD crosslinked polyacrylamide gels does not increase progressively with increasing per cent C above 10 per cent and thus, gels at equivalent molar concentrations are more restrictive above 10 per cent T than those cross linked with BIS. Furthermore, DATD crosslinking leads to short average chain lengths and pools of unpolymerized crosslinker which increase as the polymer chain lengthens. These properties are those that provide DATD gels with the practical advantages of relatively enhanced elasticity, mechanical stability and adherence to glass walls (24). Gels prepared with DATD are also reversible in that they can be solubilized by subjecting them to oxidation with periodic acid which cleaves the 1,2-diol stucture (25).

In addition, a new material AcrylAide (Marine Colloids, FMC), an olefinic agarose derivative has been developed to replace small organic crosslinkers such as BIS. This material has the advantage of allowing one to dry at 60° C without vacuum thicker, high per cent T or pore gradient gels backed on polyester films without their cracking or peeling from the support material.

There are a number of other "special property" crosslinkers available for specific purposes. O'Connell and Brady (26) reported the use of N,N'- (1,2-dihydroxyethylene) bisacrylamide (DHEBA) for casting reversible gels. The 1,2-diol structure is readily cleaved by oxidation with periodic acid. Paus (27) has reported on the use of Ethylene diacrylate (EDA), which can be cleaved by base hydrolysis. Polyethylene glycol diacrylate 200 and 400 ($PEGDA_{200}$ and $PEGDA_{400}$) were reported by Righetti *et al.* (28). An additional reversible gel has been described by Hansen *et al.*(29), which employs *N,N'* -Bisacrylcystamine (BAC) as the crosslinking agent. This agent contains a disulfide bridge cleavable by thiols. These gels appear to be particularly useful for fractionation of RNA, and are the only ones which can be liquified under very mild and almost physiological conditions.

1.3. AGAROSE

1.3.1. Background

Agarose is a purified linear galactan hydrocolloid isolated from agar or recovered directly from agar-bearing marine algae. Different agarose preparations vary significantly with respect to their physical and chemical properties. Agarose is obtained from the agar-bearing seaweeds from the class Rhodophyta, or red algae. While agar may be obtained from at least five different species, *Gelidium and Gracilaria* are the two most important commercially. Since the 1950s members of the *Gracilaria* species have emerged as important sources of agar,the starting material for agarose preparation. Methods have been developed to enhance the gel strength of *Gracilaria* agar by conversion of the 6-sulfated *L*-galactose moieties into 3,6-anhydrogalactose, a natural process carried out enzymatically in *Gelidium*.

As a result of this conversion, the derived agarose has a higher gel strength and generally lower sulfate concentration than *Gelidium* agarose (30).

Agarose is an alternating 1,3 linked β-D-galactopyranose and 1,4-linked 3,6-anhydro α-L -galactopyranose, as shown in Fig. 6.

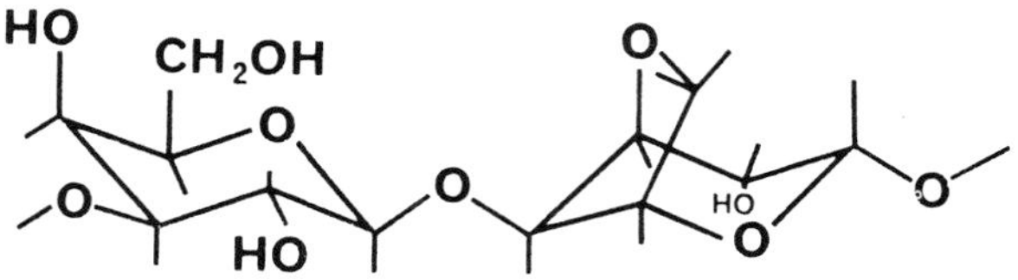

Fig. 6. Agarobiose: Basic repeating unit of agarose

The properties and types of agarose covered in this section are limited to those useful in isoelectric focusing and a brief mention of the "so-called" un-supercoiled agarose which has similar sieving properties to acrylamide (31).

Agarose *per se*, does not exist - only "many different agaroses do". These fall into the following three categories:

1. The first catagory are the agaroses graded according to the degree of substitution with negatively charged sulfate and carboxylate groups and the degree of electroendosmosis that they exhibit. Ranked in descending order of electroendosmosis, these are the Marine Colloids (FMC) products trademarked **HEEO, HE, ME** and then **LE, HGT** and **HGT(P)**, and the Litex products trademarked **HSA, HSB**,and **HSC.** These products are listed for the readers information to indicate the diversity of products; however these will not be discussed further as electrophoretic support media.
2. The second category is the so-called "Charge free, or Zero electroendosmosis agaroses", which are dicussed below.

3. The third category of agaroses are those agaroses that have been hydroxyethylated. These are capable of producing gels with decreased pore sizes relative to proteins and lowered melting points (32). The most highly derivatized species (about 10 per cent) is produced by Marine Colloids (FMC) under the trade name **SeaPrep™**. It is similar to the product of Litex produced under the tradename **Agarsieve™**. SeaPrep agarose melts at room temperature and at high concentrations it exhibits pore sizes similar to crosslinked acrylamide (31). A second of this type of agarose, hydroxyethylated at about five per cent, from Marine Colloids is produced under the tradename **SeaPlaque™**, which melts at a higher temperature and presumably corresponds to Litex **EasyPlaque™** and LS-Series.

1.3.2. Non-sieving, "charge free" agarose

The requirement of a "charge free" agarose for use in isoelectric focusing has led to the development of three commercially available agarose preparations for this purpose. All are prepared using different techniques of charge blocking or by the inclusion of additives to prevent electoendosmotic flow with its particularly disasterous effects in isoelectric focusing.

The first of these is produced by Marine Colloids (FMC Corp., Rockland ME),under the trade name of **IsoGel™**. This material is a mixture of specially prepared agarose, to which has been added 25 per cent clarified locust bean gum. The added viscosity from this material effectively inhibits electroendosmotic flow allowing isoelectric focusing on thin layer gels 0.8mm thick (33) (see Isoelectric Focusing Chapter 3, section 3.14.) .

A second preparation is produced by Pharmacia under the tradename **Agarose IEF™**. This material has been made suitable for IEF by balancing the negative charges from residual sulfate ions, presumably with quartenary ammonium groups with pH values outside of those normally used in IEF (34). However, this procedure may cause binding of the proteins with higher isoelectric points to the zwitterions present, with the consequence that any such complexes formed will have isoelectric points different from that of the native protein. However, little work has been done on any such affects in this material to document this experimentally. The user should be aware of this as a possible producer of artifacts in the apparent isoelectric point of proteins with isoelectric points above a pH of 7.5.

Besides its large pore size, allowing focusing of such large molecules as zinc glycinate human tumor marker (> 2 million daltons), as shown by Saravis *et al.* (35), Agarose also has other advantages over acrylamide: (a) it is non-toxic, while acrylamide monomer and the normally used Bis are cumulative neurotoxins; (b) it is gelled without the aid of catalysts, although, new procedures (to be released) allow this also with acrylamide; (c) it is readily compatible with crossed immuno-electrophoretic techniques; (d) it allows quick and easy staining of proteins and enzymes, as well as with appropriate high sensitivity silver staining (See Chapter 5, section 5.3.). Agarose, previously, also enjoyed the advantage of being readily dried, serving as a permanent record. However, now ultrathin-layer acrylamide gels bonded to glass or polyester films may be similarly dried and thicker gels using AcrylAide™ as a cross linking agent (see section 1.2.3.) also may be dried with excellent success to produce a permanent record.

All of these commercially available preparations may be utilized giving acceptable separation with isoelectric focusing. It should be kept in mind; however, that such unknown materials as clarified locust bean gum could have adverse effects in some, but as yet undetermined systems. On the other hand, binding artifacts may also occur in preparations where zwitterions such an ammonium have been included.

1.3.3. Agarose as a sieving support medium

The report of Nochumson *et al.* (36) and the more recent study of Buzas and Chrambach (31) have shown that agarose containing 19 per cent hydroxyethyl groups, prepared by Marine Colloids under the trade name **SeaPrep** 15/45™, exhibits sieving properties similar to cross linked polyacrylamide gels, when compared over a concentration range of 2 to 10 per cent. The 15/45 designation indicates the gelling temperature and melting temperature respectively of the material. However, there are a number of disadvantages at present to this material as compared to cross linked polyacrylamide gels: (a) the viscous solutions can not be pipetted; (b) the gels are far more mechanically labile than polyacrylamide gels; (c) also they have a tendency to melt due to Joule heat at the regulated current levels (4ma/cm² usually applied to polyacrylamide gels; (d) they frequently must be stained and destained in the cold, which leads to longer stain times and background stain problems.

The properties of this material presently are not thought to be from an un-supercoiling of the native agarose by hydroxymethylation, but due to the increased concentration of packed supercoils (37). **SeaPrep** 15/45™ has been mentioned here only to indicate that agarose may be utilized as both a nonsieving, as well as a sieving support medium. It awaits for future developments to determine the actual usefulness of this potentially interesting support medium.

1.4. References

1. Bloemendal, H., Jongkind, J. F. and Wisse, J. H. : *Chemisch Weekblad 58,* 501 (1962).
2. Ott, H. : *Prot. Biol. Fluids 10,* 305 (1963).
3. Raymond, S., and Weintraub, L. : *Science 130,* 711 (1959).
4. Allen, R. C. : in *Electrophoresis and Isoelectric focusing in Polyacrlamide Gel,* (eds.) Allen, R. C. and Maurer, H. R., de Gruyter, Berlin, p.304 (1974)
5. Allen, R. C., Popp, R. A., and Moore, D. J. : *J. Histochem. Cytochem. 13,* 249 (1965).
6. Raymond, S. and Nakamichi, M. : *Anal. Biochem. 7,* 697 (1962).
7. Ziabicki, A. : *Polymer 20,* 1373 (1979).
8. Righetti, P. G. in : Allen, R. C. and Arnaud, P. (eds.), *Electrophoresis '81* deGruyter, Berlin, p. 3 (1981)
9. Blattler, D. P., Garner, F., Van Slyke, K. and Bradley, A. : *J. Chromatogr. 64,* 147 (1972).
10. Rodbard, D. and Chrambach, A. : *Proc. Nat. Acad. Sci. USA, 65,* 970 (1970).
11. Peterson, J. I., Tipton, H. W., and Chrambach, A. : *Anal. Biochem. 62,* 274 (1974).
12. Toombs, M. P. : *Biochem. J. 96,* 119 (1965).
13. Righetti, P. G., Brost, B. C. W., and Snyder, R. S. : *J. Biophys. Methods 6,* 347 (1981).
14. Campbell, W. P., Wrigley, C. W., and Margolis, J. : *Anal. Biochem. 129,* 31 (1983).
15. Rüchel, R. and Brager, M. D. : *Anal. Biochem. 68,* 415 (1975).
16. Rüchel, R., Steere, R. L. and Erbe, E. F. : *J. Chromatogr. 166,* 563 (1978).

17. Chrambach A., Jovin, T. M., Svendsen, P. J., and Rodbard, D. : in *Methods of Protein Separation, Vol. 2,* Catsimpoolas, N., (ed.), Plenum Publishing Corp., New York, p. 54 (1976).

18. Allen, R.C. unpublished observations.

19. Righetti, P. G., Gelfi, C., and Bianchi Bossio, A. : *Electrophoresis 5-6,* 291 (1981).

20. Davis, B. J. : *Ann. N. Y. Acad. Sci. 121,* 404 (1964).

21. Brackenridge, C. J., and Bachelard, H. S. : *J. Chromatogr. 41,* 242 (1969).

22. Chrambach, A., and Rodbard, D. : *Sep. Sci. 7,* 663 (1972).

23. Maurer, H. R., and Allen, R. C. : *Clin. Chim. Acta 40,* 359 (1972).

24. Jackiw, A. : *XI International Congress Biochem.* Montreal, Abstract p. 716 (1979).

25. Anker, H. S. : *FEBS Lett. 7,* 293 (1970).

26. O'Connell, P. B. H., and Brady, C. J. : *Anal. Biochem. 76,* 63 (1976).

27. Paus, P. N. : *Anal. Biochem. 42,* 327 (1971).

28. Righetti, P. G., Tidor, G.,and Ek, K. : *J. Chromatogr. 220,* 115 (1981).

29. Hansen, J. N., Pheiffer, B. H., and Boehnert, J. A. : *Anal. Biochem. 105,* 192 (1980).

30. *The Agarose Monograph*-FMC Corp., Rockland ME (1982).

31. Buzas, Z., and Chrambach, A. : *Electrophoresis 3,* 130 (1982).

32. Cook, R. B. : *US Patent No. 4,312,739* (1982).

33. Saravis, C. A., O'Brien, M., and Zamcheck, N. : *J. Immunol. Methods 29,* 97 (1979).

34. Righetti, P. G. : in *Isoelectric Focusing : theory, methodology and applications,* (eds.), Work, T. S. and Burdon, R. H., Elsevier, Oxford, p. 165 (1983).

35. Saravis, C. A., Cunningham, C. G., Marasco, P. V., Cook, R. B., and Zamcheck, N. : in *Electrophoresis '79* Radola, B. J., (ed.), deGruyter, Berlin, p. 117 (1980).

36. Nochumson, S., Cook, R. B., and Williams, K. W. : *Elektrophorese Forum '80* Bode, München (1980).

2. POLYACRYLAMIDE GEL ELECTROPHORESIS

2.1. Charge Size Fractionation : Background

The often perplexing question asked in regard to electrophoresis is which system to use for a given separation problem. Fractionation by electrophoresis of macromolecules can be achieved effectively by exploiting the three major properties of macromolecules, namely; size, net charge and relative hydrophobicity. Polyacrylamide gel electrophoresis (PAGE) and SDS-Page, polyacrylamide gel isoelectric focusing (PAGIF) and agarose isoelectric focusing (AGIF) or in combination, provide the basic means to accomplish this. The methods serve essentially three functions: Isolation of homogeneous native macromolecular species, their physical characterization and macromolecular mapping.

All of the methods developed toward this aim have a valid place in the fractionation and separation of proteins. On the other hand, misuse of a particular method, or use of the wrong method, can lead to disasterous results. It is the aim of this chapter to aid in the choice of rational, objectively defined systems to allow the reader to achieve optimally efficient, practical fractionation conditions for any given task where these methods are appropriate.

In this chapter, the general methods for utilizing the properties of charge (mobility), size and hydrophobicity in sieving media are discussed. Separation by charge alone in non-sieving PAGIF and AGIF is dealt with separately in Chapter 3.

2.1.1. Zone electrophoresis *vs.* moving boundary electrophoresis

The aim of all electrophoretic methods in sieving media has been to achieve the maximal resolution from any complex mixture for analytical purposes, or between any two closely migrating components, in fractionation procedures. The major prerequisite toward this goal is to initiate the separation under conditions where the sample is concentrated into the thinnest possible starting zone in order to limit the effects of diffusion of protein zones during separation with its resulting decrease in resolution. The approaches to this, on the surface, relatively straight forward problem have led to literally hundreds of theoretical papers since 1964 espousing one or another procedure to accomplish this goal. The two major directions have been in the development and improvement of continuous and discontinuous buffer systems with the result of no universal application for either system. The major contribution, unfortunately, has been the confusion caused to the user who must attempt to select the appropriate system for his or her particular separation problem.

With respect to the distribution of electrolytes, PAGE can be divided into two general categories: zone and moving boundary electrophoresis. In the former the ionic environment is homogeneous throughout the system, consisting of a single species of ion and counterion. In the latter, the systems require initially at least two different ions to be present in addition to a counter ion common to both systems. During the electrophoretic process the Kohlrausch boundary, formed at the interface of the two ion layers, may migrate either toward the anode or the cathode, depending on the buffer ions chosen and the pH of the system.

The earlier use of PAGE as an analytical method for the separation of complex mixtures of proteins has, in large part, given way to its more prevelant present use for size fractionation, preparative studies and to define the characteristics of a given protein. However, the question of whether one should select continuous zone electrophoresis (CZE), or a discontinuous multizonal electrophoresis (MZE) system is normally not addressed. When it is addressed, it is usually found as part of a highly theoretical paper or review, that all too often appears to have been overlooked. It is, perhaps, the overly ambitious task of this chapter to try to bring some order out of this chaos.

Chrambach *et al.* (1) have succinctly stated the problem as follows:

> "There is no necessity to use MZE in the relatively uncommon cases where the species of interest is available in high enough concentration (1mg/ml or more) to allow for the formation of thin sample zones by concentration of sample (preferably applied at low ionic strength) on the surface of a restrictive resolving gel. If concentration by MZE is not needed, continuous zone electrophoresis (CZE) can be carried out in a single buffer (identical in both buffer reservoirs and in the gel) at any pH with characterization of bands in terms of absolute mobility (cm^2/sec/volt) (Rodbard and Chrambach 1974 (2); Morris and Morris, 1971 (3). CZE should also be used where ionic additions to an MZE system would seriously perturb its operative properties; this entails sacrificing the convenience of automatic, electrophoretic sample concentration."

These admonitions are as valid in the overall separation process approach today, as when they were written. Although, the recent advent of 200 fold more sensitive silver staining (discussed in Chapter 5 - section 5.3), would probably lower this limit of 1 mg/ml to 50 - 100 µgm/ml. In any event, an understanding of the CZE systems, their capabilities and limitations is necessary in the selection between CZE and MZE to select the appropriate separation procedure for charge and size separations.

2.1.1.1. CZE Electrophoresis

Hjerten *et al.* (4), applying the method introduced by Haglund and Tiselius (5), demonstrated that by decreasing the conductance of the sample (accomplished by dialysis of the sample against water, or merely by dilution with a buffer of one-half to one-fifth the ionic strength of the separating gel buffer) that sample concentration and a thin starting zone could be obtained and that resolution in such a system was comparable to that achieved by discontinuous boundary zone sharpening, as shown in Fig. 1.

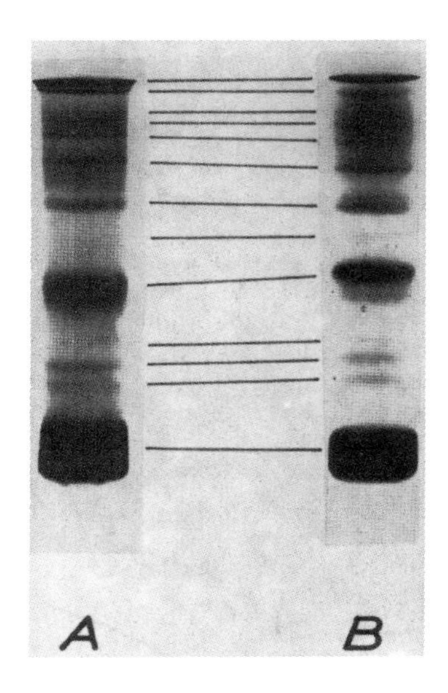

Fig. 1. Electrophoresis of human serum in a polyacrylamide gel: A. In a discontinuous buffer system according to Ornstein and Davis (6), B. In a continuous buffer system at the running pH of system A. Horizontal lines indicate zones presumed to be identical. As seen, the discontinuous and continuous buffer systems give similar patterns. (from Hjerten *et al.*) (4)

This procedure demonstated that similar results could be obtained with both CZE and MZE, but in neither case was the full advantage of the sieving properties of acrylamide utilized.

Slater in 1965 (7) reported that certain plasma and cell proteins could be resolved better in polyacrylamide gel gradients. Slater (8) pointed out, that the migration rate of proteins in a linear gradient gel decrease with time; eventually a particular protein will reach "its" pore limit and stop. By such a process, a protein zone is also automatically sharpened, with a resultant increase in resolution. Pore limit electrophoresis, thus, provides a valuable means for separating by size fractionation a protein mixture of widely differing sizes. Margolis and Kendrick (9) performed a number of studies also by continuous linear and curvilinear pore gradient electrophoresis, an example of which is shown in Fig. 2.

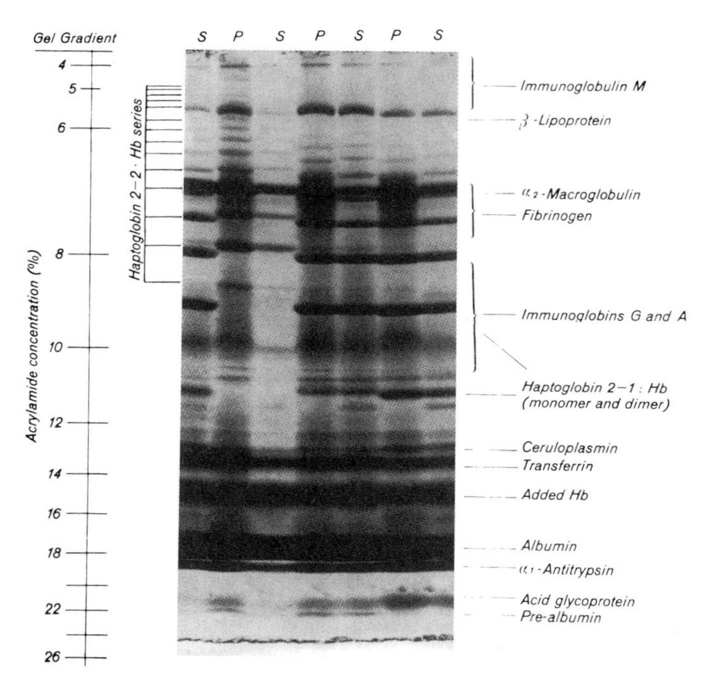

Fig. 2. Electrophoresis of human plasma (P) and serum (S) proteins in a 4 - 26 per cent concave gel gradient. (from Margolis and Kendrick) (9).

With CZE, any charged macromolecule can be separated with a high degree of resolution. Charge and size may be utilized in a homogeneous gel to achieve the separation pattern on the one hand, while, size alone is the determining factor of the gradient separation pattern. In the former case for example, α_1-antitrypsin is found cathodal to the albumin in a 7.5% homogeneous gel since it has a lower mobility than albumin. In the pore limiting gradient gel, sometimes referred to as a "pore stop gel", the effect of size becomes predominant and the smaller 54,000d α_1-antitrypsin migrates further into the pore gradient gel than the larger 68,000d albumin, despite its lower mobility. Therefore, the positional relationship of particular proteins in a pattern can be affected by the pore size of the gel selected in homogeneous systems.

2.1.2. Discontinuous and Multizonal Electrophoresis Systems

2.1.2.1. Disc Electrophoresis

In 1957 Poulik (10) showed that the resolving power of starch gel electrophoresis could be improved significantly with discontinuous buffer systems, when a borate buffer as the electrode medium, but a citrate buffer in the the starch gel was used. He observed a boundary migrating toward the anode, which sharpened the protein zones as it overtook and passed them and attributed this to *a discontinuity in voltage gradient* . It remained for Ornstein (11) to recognize that the affect observed by Poulik was due to the Kohlrausch regulating function that first described moving boundaries in multiple ionic systems in 1897 (12). This, and his own observation, that in borate buffered polyacrylamide gels the breakdown product of the high concentrations of persulfate utilized to catalyze the gel (sulfate) formed a moving boundary in the gels, led to the development of the *disc* method of Ornstein and Davis (6).

By selecting a weak acid as a trailing ion and operational shifts in pH, the effective mobilities $m\alpha$ of all ionic species could be arranged so that a sample could be sandwiched between a leading and trailing ion to concentrate it and to provide a thin starting zone in the following manner:

$$m\alpha_{\text{leading ions}} > m\alpha_{\text{proteins}} > m\alpha_{\text{trailing ions}}$$

As a note of historical interest, Ornstein (11) found in retrospect that the studies of Kendall *et al.* (13) using moving boundary methods had resulted in the sucessful separation of traces of Radium from Mesothorium I and Barium and also separated a number of rare earth ions one from another (14,15) all on agar-agar gel columns. In fact, in an article in *Science* (16) in 1928 Kendall suggested the use of the moving boundary for protein separation.
A schematic presentation of the *disc* system is shown in Fig. 3.

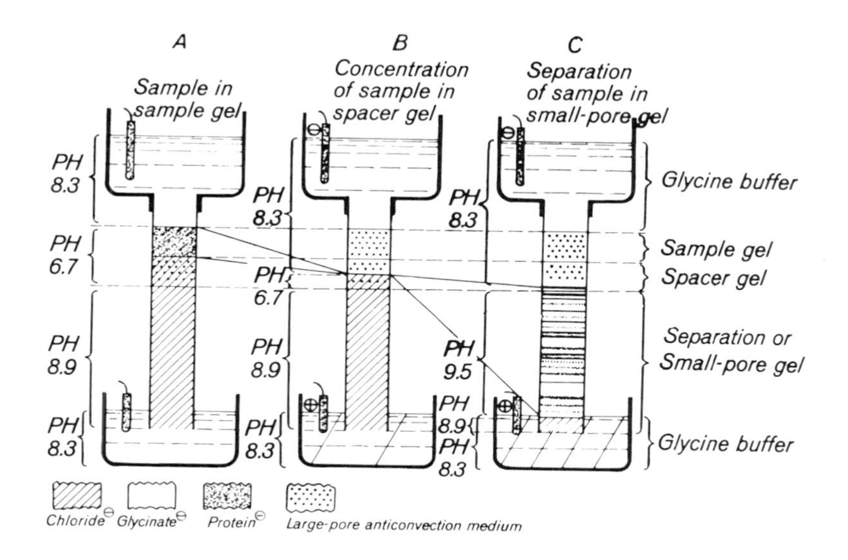

Fig. 3. Disc electrophoresis (from Ornstein) (11)

The three phases of the Ornstein system and their interrelationship are given in Table 1. below to clarify the operational pH shift, concentration, steady state stacking and sieving steps inherant in the system (Note: The alignment of constituents in the stack is in order of net mobility, not net charge. Monovalent acids and bases with the same net charge of 1 may differ in mobility by a factor of 3 - 4.).

Table 1. Operational steps in the Ornstein "Disc" System

1. Sample Gel.	Under the conditions shown in Fig. 3. A, when current is applied to the system; the glycine ion, from a weak base, at pH 8.3 is only partially dissociated and has a mobility less than the slowest component of the serum sample (γ-globulin). The chloride ion, from a strong acid is completely dissociated and has a mobility, based on its transfer number, faster than any other charged components in the system. Thus, the sample components are sandwiched between the leading chloride ion and the trailing glycine ion and aligned in the order of their mobility. The sample gel simply forms an anticonvective medium to assure that all of the sample components are included in the sandwich.
2. Stacking Gel	In Fig. 3. B. The sandwich enters the spacer ("stacking") gel and the stack of ions and proteins becomes compressed due to the higher voltage gradient behind the leading edge of the trailing ion, thus, a tight stack is formed; which also serves to concentrate the sample components.
3. Separating Gel	At the interface of the spacer and separating gel there is an operational shift in the pH of the support medium (increase) , which causes the glycinate ion to be further dissociated, thus increasing its mobility. At this juncture, the glycine ion over takes the sample components and runs just behind the chloride ion.
4. Sieving Gel	Also at this juncture the pore size of the gel becomes restrictive and the protein components begin to separate from one another both by charge and size. Once the ion boundaries have passed ahead of the proteins the separation takes place as in zonal electrophoresis with one exception. That is that there remains a voltage gradient across the gel that increases, in this case, from the boundary to the cathode. This has the consequence that any protein component continues to find itself in an electrochemical microenvironment where the voltage gradient is lower ahead and higher behind. This serves to continue to keep each component tightly compressed. In fact from the instant the glycine ion overtakes all of the protein components the slowest charged protein becomes the trailing ion.

The Ornstein system was designed for the separation of serum proteins and was a compromise of pH and gel pore size to obtain optimal separation. This system has led to considerable confusion, since Ornstein failed to state the operative conditions (pH and ionic strength differences) in this system to which the sample was actually exposed to. A number of subsequent variations of this system were developed in both cylindrical tube and slab systems to utilize the moving boundary sharpening effect, but without the sample gel , since some sample components may be denatured by the free radical formation that occurs during the photopolymerization process (17) and from the discontinuities in pH gradient to which the sample was subjected (4). Some consequences of sample gel deletion are depicted in Fig. 4.

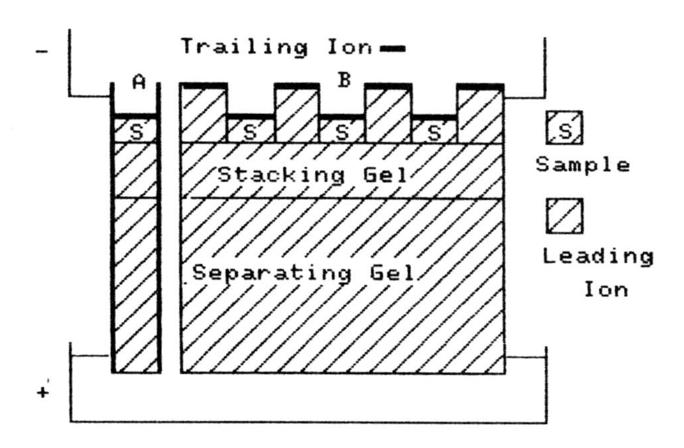

Fig. 4. Initial boundary conditions when current is applied: Tubes, A and Slab gels, B; where no sample gel is used.

In such instances, when the sample gel is deleted the boundary forms at the interface of the sucrose - sample mixture and moves rapidly through causing convective currents with the result that part of the sample may not be in the initial stack; thus, causing vertical streaking to occur subsequently from non-stacked proteins. There are additional differences between tube gels, depicted in panel A. and in slab gels, panel B. In the latter, not only will similar convection occur, but also the boundary forms in a stepwise fashion due to the gel fingers forming

forming the wells, normally extending well above the level of the sample. This can lead to pinching or flaring of the pattern if sufficient ionic strength differences are present between the sample and the adjacent gel divider between sample lanes. It will also tend to spread the sample track due to the initial uneven boundary, whether the well bottoms are cast "optically flat" or not. (see gel casting section 2.6.)

2.1.2.2. Discontinuous electrophoresis at a constant pH

There are a number of problems associated with the practical application of the *disc* system when attempts were made to extend its application to other separation problems, particularly enzymes: First, Hjerten *et al.* (4) discussed the problems that discontinuities in the pH of the system could have on the sample in the Ornstein system. Second, Allen *et al.* (17) found that in this system 15 to 25 per cent of mouse non-specific esterase activity was destroyed by polymerizing the sample in the sample gel, when compared to duplicate samples placed only in buffered sucrose and covered by a cap of sample gel. A third major shortcoming of the Ornstein system was that the pH limits of the sample and stacking gel were chosen to include γ globulin, with the lowest mobility and albumin with the fastest mobility, in the initial stack. The operational pH required in this system (pH 8.3) may lay close to or below the isoelectric point of some sample components causing them to be excluded from the stack, or even to remain in the sample gel. A good example of latter is LDH 5 with an isoelectric point above pH 8.3. This results in its remaining, equally distributed throughout the sample gel at the completion of the separation, since the pH of the operative stack was insufficient to impart a charge to the enzyme (18).

Allen *et al.* (19) and Allen and Moore (20), utilized the Kohlrausch function to provide the delayed sharpening effect noted by Poulik (10) and the conductivity shift of Hjerten *et al.* (4) in conjunction, in order to develop a more versatile and a less deleterious system for enzyme separations. By utilizing a 1.8 cm cap of gel above the sample in buffered sucrose, both containing the leading ion, the boundary was prevented from reaching the proteins until they had migrated into the gel. A two phase system at a constant pH was developed to produce thin starting zones, with concentration by a *conductivity shift* followed by *zone sharpening by a moving boundary* in sequential order. This was followed by zone stacking and unstacking similar to the Ornstein system. A schematic presentation of this system is shown in Fig. 5. below.

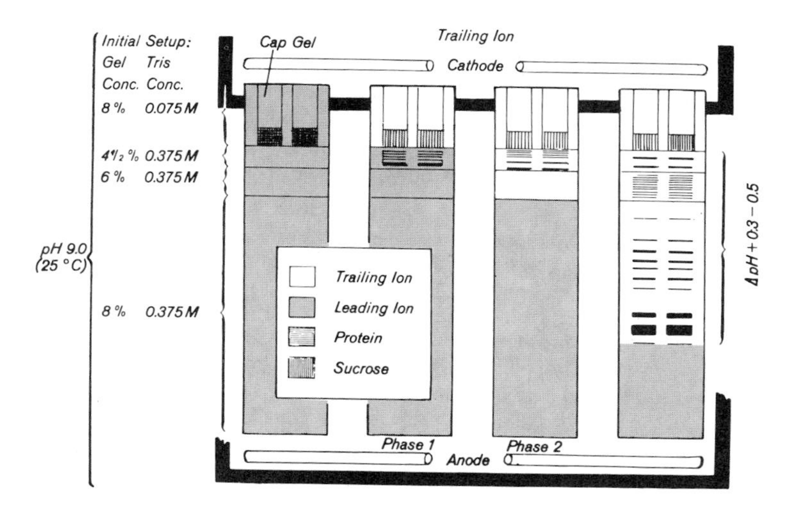

Fig. 5. Setup and operational steps using a conductivity shift to partially concentrate the sample in a stacking gel followed by a moving boundary to stack and then unstack and separate the proteins. (from Allen *et al.*) (20)

The three phases of the discontinuous buffer system at a continuous pH are given in Table 2. below. The first two phases are divided into parts a and b for clarity of presentation only, since they are ocurring simultaneously.

Table. 2. Operational steps in discontinuous buffers at a continuous pH systems

1. Phase 1a.	In the first phase the sample, nominally pH 9, migrates into the first (more open pore) layer of the step gradient separating gel. A five-fold shift in conductance between the sample in the leading ion buffer and the gel are employed to allow rapid migration into the gel with little or no convection and at the same time produce sample concentration.
Phase 1b.	The moving boundary is formed at the interface of the upper buffer and the top of the cap gel. The cap gel is kept sufficiently deep (normally 1.5 -1.8 cm) to allow the sample to enter the first gel layer completely before the ion boundary passes through the now empty sucrose buffer solution. This phase is carried out at low current or power levels.
Phase 2a.	The boundary now overtakes each of the partially separated proteins and in order of their mobility sharpens each one as it passes. This is accomplished since the rear of each protein zone is momentarily in a higher voltage gradient than the front edge, thus, each zone is voltage gradient sharpened by a conductivity jump and then stacks on top of the boundary sequentially in order of both mobility and size of larger macromolecules that may be sieved by a the 4.5 % gel.
Phase 2b.	Once the boundary has passed through all of the protein components, the protein with the lowest mobility forms the *trailing ion and steady state stacking occurs for all components not sieved.*
Phase 3.	When the boundary has passed through all the protein components the power to the system is increased and the proteins unstack and separate as in free zone electrophoresis, but in contrast to the *disc* system they also migrate into an increasingly smaller pore size gel. All separations with the exception of SDS-PAGE are carried out at 4-8ºC.

A comparison between the Ornrnstein (11) glycine - chloride system and that of Allen *et al.* (19) with borate - sulfate at a continuous pH of 9.0 is shown in Figs. 6 and 7.

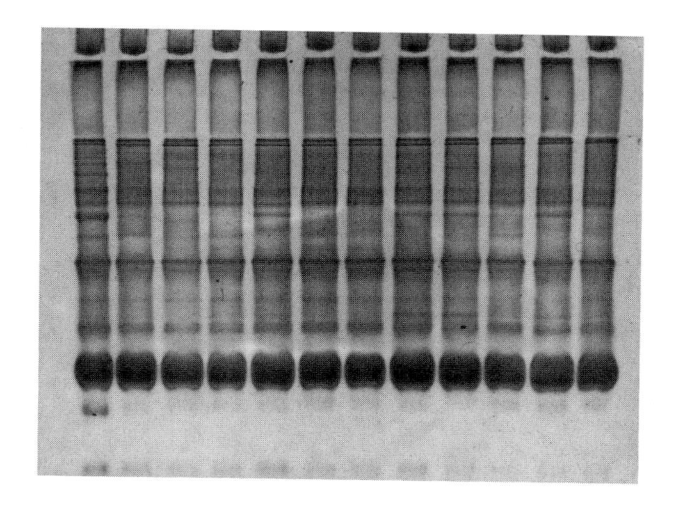

Fig. 6. Human serum proteins separated at a constant current of 60 ma in a 3mm thick gel slab using the Ornstein system. The gel was fixed in TCA and stained with Coomassie Brilliant Blue R 250 (21). (from Allen and Maurer) (22).

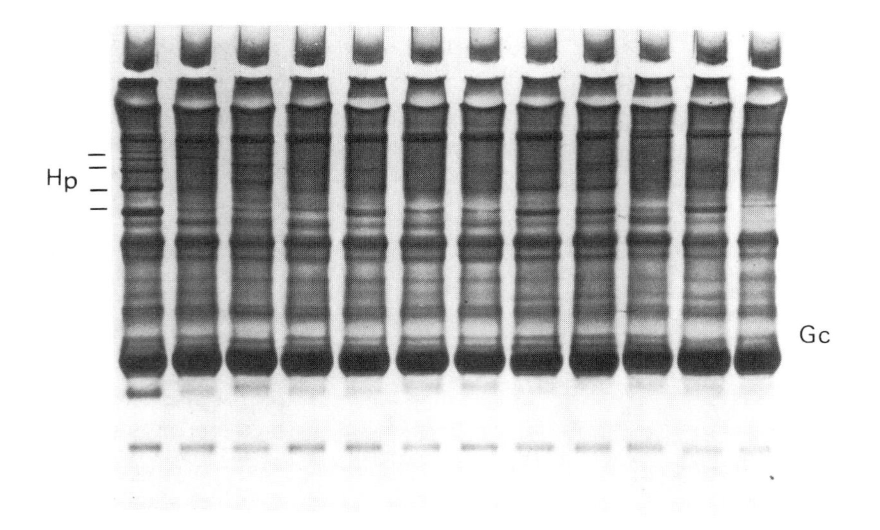

Fig. 7. Same samples as in Fig. 6. under same current, temperature and run length, but with a 3.5 - 6 - 8 per cent T step-gradient gel. The buffer system was borate - sulfate at pH 9.0. (from Maurer and Allen) (22).

As may be seen in Fig. 7, the separation patterns differ considerably from those in Fig. 6. While sieving may play a small role in the difference, the major contribution to the increase in resolution of the Gc proteins and the sharper albumin and prealbumin bands is due to the different MZE system. Protein zones in the post-ceruloplasmin region are also resolved better as evidenced by the various haptoglobin phenotypes.

The buffer system used in the development of the last system was selected as a compromise to achieve optimal separation of complex mixtures of macromolecules, in contrast to the mathematical derivation involved in the ion and pH selection of the Ornstein system. The major reason for this is the mathematics involved in similar calculations for multivalent ions require an infinite series expansion, presently not soluble. However, certain empirical relationships may be obtained based on the practical application of the *disc* system theory and the of multizonal electrophoresis discussed in a later section. These effects are shown to aquaint the reader with the practical separation effects of various buffer ion combinations at a single pH and temperature on the separaton profile of human serum as examples shown in Fig. 8. These are not necessarily recommended systems, but are presented only as a demonstration of apparent boundary mobility effects. The ratio of the boundary mobility to that of the albumin front, shown in Table 3., provides an index of overall resolution in a homogeneous or simple step gradient gel system.

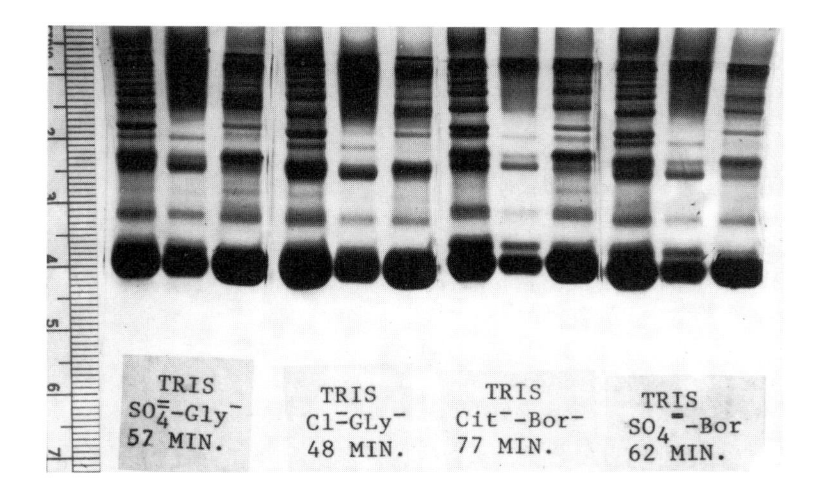

Fig. 8. Normal and abnormal human serum proteins separated at a constant current of 90 ma in a 3mm thick 4.5 - 6 - 8 % T step gradient slab gels stained with Amido Black 10 B .(from Maurer and Allen) (22).

Table 3. Relation of the boundary velocity to the velocity of the albumin front for various leading-trailing ion combinations at 75 ma constant current and 4°C.

Trailing Ion	Leading ion	Boundary mm/min.	Albumin mm/ min.	Ratio
Glycine	Chloride	1.91	0.79	2.17
Glycine	Sulfate	1.08	0.67	1.74
Glycine	Citrate	0.74	0.53	1.40
Borate	Chloride	1.07	0.71	1.51
Borate	Sulphate	0.95	0.62	1.53
Borate	Citrate	0.66	0.50	1.32

In Fig. 8, it may be seen that the higher the ratio of the velocity of the boundary relative to that of albumin front, the less the overall resolution achieved, particularly in the post albumin region. One explanation offered for this effect, is that the voltage gradient increases from the boundary to the sample application point. Thus, Joule heating and diffusion increase in the those systems with the highest boundary mobilities. A second effect is that the proteins will unstack earlier in faster boundary systems, which tends to cause zone broadening.However, it also follows that the separation time is shorter in such systems as shown in Fig. 8. For example, the albumin migrates 4 cm in 48 min in the fastest glycine - chloride system, while requiring 77 min in the slowest borate - citrate system; yet overall resolution is superior in the latter.

2.1.2.3. Multizonal Electrophoresis (MZE)

The foregoing two sections have described the *disc* method specifically tailored for the separation of serum proteins and a more general approach, empirically designed, for the separation of complex mixtures of enzymes. These systems are, through necessity, compromises to achieve optimal over all resolution of complex mixtures and do not address a number of other applications that best can be accomplished with MZE.

Today, moving boundary electrophoresis on gels is useful for four purposes: 1) As an automatic concentration method to produce ultrathin starting zones of proteins, or other charged macromolecules, for analytical and preparative resolving gels; 2) to provide a sharply defined reference zone in the resolving zone for R_f measurements; 3) for the preparative electrophoretic extraction of proteins from gel slices; 4) for high load preparative "charge fractionations" of relatively simple mixtures of proteins. It requires buffer systems at various pH values which are capable of setting up moving boundaries with numerically known leading and trailing ion mobilities. Since protein migration is slow compared with that of buffer ions, these moving boundaries have to have low displacement rates if they are to concentrate and align the proteins in order of mobility.

This problem was addressed first in the middle 1960s by Jovin and resulted in the appearence, several years later, of the *Multiphasic Buffer Systems Output* by Jovin *et al.* (23). This heroic presentation of some 4269 buffer systems, covering the pH range of 2.5 to 11 in 0.5 pH units of systems for 0° C and 25° C, for positively and negatively charged ions was made available in computer output form or by microfische. Subsequently Jovin (24) and Everaerts *et al.* (25) published theoretical treatments that were more comprehensive and correct than the original Ornstein theory of discontinuous electrophoresis (11) and with the additional advantage that they were available as computer programs to make use of their potentially wide application. More recently Schafer-Nielsen and Svendsen (26) have used a graphic approach to define the mobilities of leading and trailing ion buffer constituents across a moving boundary as a funtion of pH.

All of these programs and early studies of Svensson (now Rilbe) (27,28) and Longsworth and his coworkers (29), steady-state stacking and *disc* electrophoresis of Ornstein (11), *multiphasic zone electrophoresis* (MZE) of Jovin (24), *isotachophoresis* of Everaerts *et al.* (25), *displacement electrophoresis* of Martin and Everaerts (30) and Hjerten *et al.* (31) and , last but not least, *snow electrophoresis, telescope electrophoresis and steady state stacking electrophoresis* of Schafer-Nielsen and Svendsen (26) all describe identical electrophoretic phenomena. Even isoelectric focusing, discussed in Chapter 3, appears to be a special case of moving boundary electrophoresis in which the proton and hydroxyl ion are the sole common ions (32,33).

The use of these systems as a practical tool for the biologist and biochemist, just has not taken place. Presumably the reasons for this are fivefold: 1)Confusion existed about which of the various forms and techniques of moving boundary electrophoresis applied to what; 2) the variety of names for essentially the same technique; 3) only a few of the described buffer systems exhibited sufficiently low moving boundary displacement rates to allow slowly migrating proteins to migrate within the moving boundary and 5) it proved difficult and expensive for potential users to interpret and utilize the tapes or microfische systems available.

Chrambach and Jovin (38) have recently simplified the nomenclature and have reduced the number of buffer systems from over four thousand to a manageable 19. The reader is referred to this article for the selection of the appropriate buffer system for accomplishing fractionation problems by charge and size.

2.2. The Ferguson Plot

The Ferguson plot is a most useful tool and the data may be obtained by either continuous or discontinuous PAGE. The importance of its use in multizonal electrophoresis (MZE) systems for selection of the best gel pore size, discussed in the last section, is as critical as is its more conventional use for molecular weight determination. We have thus, selected to present the concept at this point in order that these two sections may be viewed together.

The retardation effect on a randomly shaped particle, passing through a random meshwork of inert fibers, is a function of fiber length, fiber radius and particle radius as defined by the Ogston theory (39). The Ogston theory predicts that the progression of a particle in gel electrophoresis may be expressed as the antilog of its relative mobility in a gel, M/M_0 is an exponential function of the length of the gel fiber (l) and the surface of the particle passing through the gel (S). If the gel is of the random fiber type, *i.e.* if it consists of very long strands of polymer with a neglible number of fiber termini (see Fig. 1., Chapter 1). The analogous predictions for "point gel", *i.e.* gels consisting of fibers so short that their length may be neglected, or that in the case of a planar gel is shown in Fig. 9.

	Standard curve	Protein	Gel
K_R vs.	$\bar{R}^3$	Globular	Point
	$\bar{R}^2_{/}$ (MW) $^{2/3}$	Globular	Fiber
	$\bar{R}$	Globular	Plane
	$\bar{R}^3$ (MW)	Random coil	Fiber

Fig. 9. Standard curves for molecular weight determination for various combinations of macromolecular parameters and gel types. (from Chrambach) (40)

Applying this statistical model of a molecule passing electrophoretically through a gel and subjected to retardation, which increases in proportion to a reduction in pore size (increased monomer Concentration) predicts the "Ferguson plot": Log $M/M_0 = K_R T$.

Reproducible R_f measurements, applied to several gel concentrations (minimally 3 and optimally 7), yields the Ferguson plot as shown in Fig. 10.

It forms the basis of "quantitative PAGE" in view of the fact, as derived from the Ogston theory and well verified empirically (2), that its slope, the retardation coefficient **K_R** and its y-intercept (Y_0) are measures of molecular size and net charge respectively. In practice, the construction of the Ferguson plots of macromolecules of interest and of its contaminants which are to be separated, as well as for the purposes of physical characterization of the macromolecule, and of pore size optimization of the gel may be accomplished automatically, by means of a suitable computer program. Such a program, *PAGE-PACK*, has been written by Rodbard and Chrambach (2) and is shown in Fig. 11.

The computer output obtained by the PAGE-PACK program, on the basis of of the **R_f** values of 3-7 gel concentrations gives the following: (1) the Ferguson plot, **K_R** and Y_0; (2) the joint 95 % confidence envelopes of k_R and Y_0, by which a molecular indentity (in terms of molecular size and net charge) is defined for the buffer system in question and which serves as the perferred method for settling questions of molecular identity or non-identity between components; (3) a translation of the unconventional experimental parameters such as molecular size and net charge, **K_R** and Y_0, into conventional parameters such as molecular radius and weight, free electrophoretic mobility and molecular valence (net protons/molecule); (4) concise definition of the optimal gel concentration for the separation of the species of interest from any selected contaminant species. A complete explanation of the Ferguson plot and the variables that must be carefully controlled to achieve accurate plot data are give in reference (2). This should be carefully read before attempting to set up a Ferguson series and to extrapolate data to the PAGE-PACK program.

These parameters will also provide an indication if two closely related proteins can best be separated by size, size and charge or by charge alone, the latter of which is best done by isoelectric focusing.

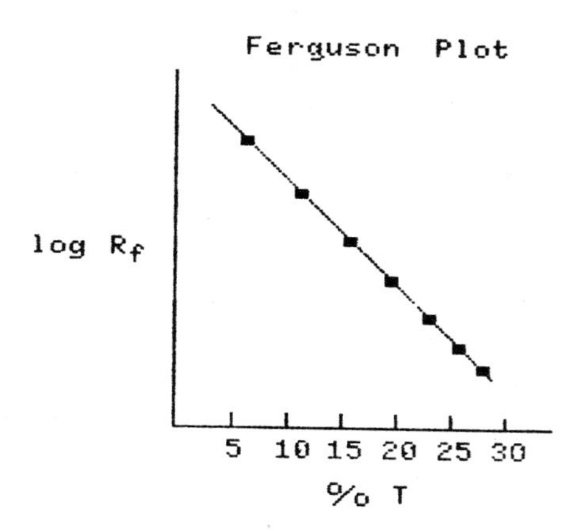

Fig. 10. Plot of mobility of an arbitrary protein against per cent T

PAGE - PACK

	Program	Input	Output
I	RFT1	R_f (%T)$_n$	K_R Y_o
II	PLOTRUN	K_R of X of Standards	$\bar{R}$, MW
III	CHARGE	Y_o, $\bar{R}$	M_o, U
IV	T-OPT	$[K_R, Y_o, \bar{R}]_A$ $[K_R, Y_o, \bar{R}]_B$	% T_{opt}

Fig. 11. PAGE-PACK program inputs and outputs (from Rodbard) (2)

2.3. Molecular Weight Determination

2.3.1. Native proteins

It follows from the previous section that molecular weight determination of native proteins may be obtained from the Ferguson plot wherein the effect of charge is eliminated in that the slope, K_R (retardation coefficient), is a measure only of molecular size.. The work of Hedrick and Smith (41) demonstated that there is a linear relationship between K_R and the molecular weight of native proteins so that, by using a standard series of known molecular weight native proteins, one may determine the molecular weight of any sample protein simply by determining its K_R and then referring to a standard curve. Unfortunately, the problem of molecular weight determination with native proteins is that it is only valid if the standard proteins have the same shape, degree of hydration and partial specific volume. Many improvements and modifications of this technique to more accurately define the molecular weights of native proteins have been made since this first report. While these authors showed that proteins in the range of 50,000 - 700,000d were correlated linearly, Gonenne and Lebowitz (42) observed a non-linearity for proteins with molecular weights < 50,000d. Rothe and Purkhandaba (43,44) recently collated the various refinements in methods for this purpose, which are summarized in the following three tables adapted from (43):

Table 4. Methods for the determination of molecular weights of native proteins by the use of homogeneous gel systems

1st Derivative	2nd Derivative	MW x 10^3	Buffer System	Ref.
log R_f *vs.* % T and determination of the slopes of the regression lines = K_R	K_R *vs.* MW	50-500 ±5 %	Multiphasic system: Asparagine-Cl, large pore gel Imidazole-Cl	(40)
log R_f *vs.* % T and determination of the slope of the regression lines = K_R	K_R *vs.* log MW	10-50 ±5 %	Buffer and gel system according to Ornstein (11), but omitting a large pore gel	(42)
log R_f *vs.* % T and determination of the slopes of the regression lines = K_R	$\sqrt{K_R}$ *vs.* $\bar{R}$ ($\bar{R}$ = radius of a weight-equivalent sphere)	40-450	Several multiphasic buffer systems at pH: 10.2, 8.88,7.8 and 5.5	(45)
log R_f *vs* % T and determination of the points of intersection of the regression lines with the log T-axis = T_{LIM}	R_S *vs.* $1/T_{LIM}$ (R_S = Stokes radius)	36-2300	Multiphasic buffer system of Ornstein (11)	(46)

Table 5. One-step methods used in the determination of molecular weights of native proteins in gradient gel electrophoresis

Derivation	MW x 10^3	Buffer System	Ref.
log MW *vs.* D or log MW *vs.* 3 log D (D = migration distance)	50 – 200 or 100 – 400	gel buffer 0.35 **M** Tris-HCl, pH 8.9 electrode buffer 0.06 **M** Tris + 0.4 **M** glycine, pH 8.3	(47)
log MW *vs.* D (D = migration distance	50 – 300	Tris-Borate pH 8.3	(48)
log MW *vs.* R_f	67 – 1300	Tris-Borate-EDTA pH 8.4	(49)
log MW *vs.*√ % T or√ D (% T = total monomer - co-monomer concentration)	43 – 670	0.1 **M** Tris-HCl pH 9.0 as gel and electrode buffer	(44)

Table 6. Two-step methods used in the determination of molecular weights of native proteins in gradient gel PAGE

1st Derivative	2nd Derivative	MW x 10^3	Buffer System	Ref.
D *vs.* t (D = migration distance t = volthours	log D *vs.* log MW	30 - 460	Tris-glycine pH 8.3	(50)
log v *vs.* % T or log R_f *vs.* % T and determination of the slopes of the regression lines = K_R (v = distance migrated per time: cm/h)	$\sqrt{K_R}$ *vs.* $\bar{R}$ ($\bar{R}$ = radius of a weight equivalent sphere)	40 - 450	Several different discontinuous buffer systems at pH 10.2, 8.88 and 7.8	(51)
$\sqrt{t}$ *vs.* D and determination of the slopes of the regression lines = a; electrophoretic run time = t and migration distance = D	log MW *vs.* a	20 - 950	Tris - Borate - EDTA, pH 8.2, or: Veronal - Tris, pH 9.8, or Phosphate, pH 7.2	(52)

2.3.2. Denatured Protein SDS-PAGE

The pioneering work of Shapiro *et al.* (53) and Maizel (54) , who observed that thiol-reduced (normally mercaptoethanol or dithiothreitol) proteins when heated with sodium dodecyl sulfate (SDS) disaggregated into protomers whose molecular mobilities were inversely related to the logarithm of their molecular weight when measured in a sieving polyacrylamide gel medium, led to the extensive use of this technique for the determination of molecular weights and for the separation of proteins and dissociated polypeptide as random coils by pure size isomerism. Since, this first report, the number of publications on molecular weight determination by SDS-PAGE has been legion. As in the case of native proteins, both CZE and MZE systems have been used with both homogeneous and pore gradient gels. However, these systems behave somewhat differently in the presence of SDS which has undoubtedly contributed to the confusion of which system for which purpose.

2.3.2.1. Basic requirements

A basic consideration in SDS-Page is the SDS itself. Many sources of this material contain considerable amount of 14 and 16 carbon chains in addition to the 12 carbon material. The former may introduce a number of undesireable affects when present. Reeder *et al.* (55) have reported that SDS less than 98 per cent pure gave streaked patterns with often ill defined bands when separated under identical conditions. This also is particularly true with the more recent techniques where the proteins or polypeptide are "renatured" in an additional step following separation under dissociating and denaturing conditons. The longer carbon chains may be irreverably bound causing loss of ability to "renature", for example, enzyme systems (56) A second, and often overlooked problem, lies in conjugated proteins (glyco-, lipo-) can not be saturated with SDS, since their non-proteinaceous moieties do not react with SDS. Furthermore, some proteins may not be completely saturated after reaction for five minutes at 100°C to the same degree (1.4 g SDS/ g of protein) as are the molecular weight standards. An der Lan *et al.* (57) have reported that filamentous hemagglutinin requires 30 min at 100 °C to achieve complete dissociation. Assuming saturation of the protein, there are two further sources of possible error: Nonlogarithmic interpolation between the migration distance of the standards, and failure to take into account the sigmoidal, rather than linear, nature of the relation between log MW and migration distance (58). Thus, in homogeneous gels it is important to ascertain that the selected gel concentration lies within the central, approximately linear segment of the sigmoidal standard curve.

Mercaptoethanol as a reducing agent may also cause problems in the presence of urea in the system. Shah (59) has reported that mercaptoethanol influences the apparent molecular weight of Jojob proteins in the presence of urea, thus, the substitution of dithiothreitol, which is 100 times more active at the same concentration as mercaptoethanol, may be desireable to use where urea is also present.

2.3.2.2. Buffer systems

MZE buffer systems have taken predominance in SDS-Page, with the Laemmli (60) technique being the most popular. This is essentially the Ornstein (11) procedure with the additional inclusion of SDS. However, since the mobility of the sulfate ion is independent of pH over the range of 3 - 11, there is no requirement to include an operational pH shift in the system. Presumably all protein and subunit species have a uniform charge and will automatically stack. Also,the observation by Wykoff *et al.* (61) that SDS migrates with a mobility higher than SDS-protein complexes in a restrictive gel, indicates that it necessarily then will overtake the zones of protein in the resolving gel if present in the sample and upper buffer. Therefore, it is unnecessary to add SDS to the stacking or separating gel. Omission of SDS has the further advantage of ensuring that conditions of polymerization elaborated for the non-SDS case remain applicable to SDS-PAGE and that stack broadening with resultant loss of resolution is minimized.

Neville (62) developed a procedure in homogeneous gels in conjunction with a borate - sulfate MZE buffer system for stacking and unstacking SDS-protein complexes over a range of 2,300 to 320,000 daltons, which provided high resolution fractionation. This system was one of the series of 4269 multiphasic buffers calculated by the theory of Jovin *et al.* (23). As a matter of historical interest, chloride ion included in the lower gel buffer (in addition to sulphate) plays no role in the electrochemistry of this system and was included due to the convention used in the original Jovin system (63). Neville's system was developed to stack all SDS-saturated proteins while leaving behind unstacked all partially saturated SDS-protein complexes. This system is of interest, since it was the first MZE system specifically tailored for SDS-protein complex stacking (63) ; although, the inclusion of the operational pH shift was not necessary in light of later findings. With the exception of the operational pH shift, this system is almost identical to the sulfate-borate system reported earlier by Allen *et al.* (19) which formed the basis of the ORTEC Slab system. A example of the borate-sulfate SDS-MZE system at a continuous pH is shown in Fig. 12.

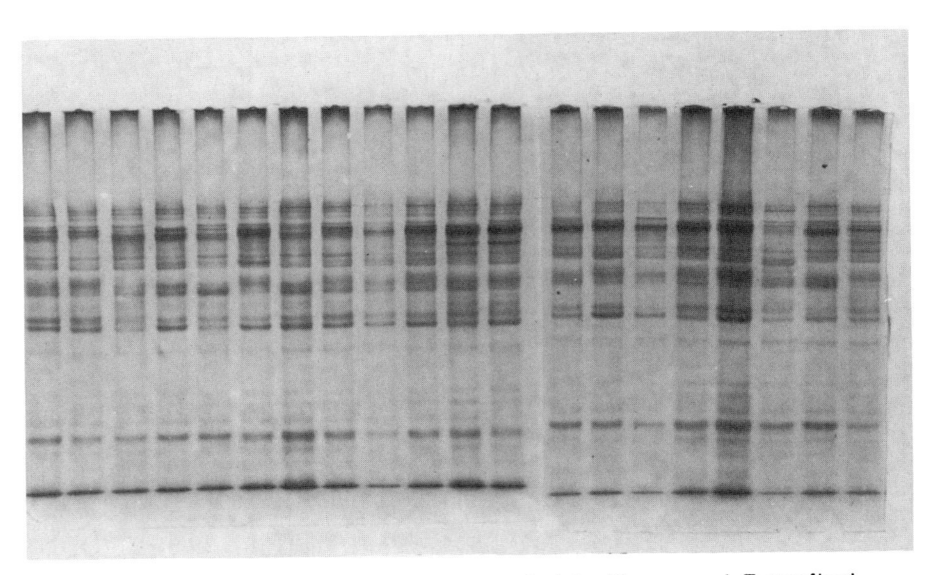

Fig. 12. Alfalfa seed proteins separated in 12 per cent T gradient with constant pulsed power. Various phenotypes are indicated by "pure size isomerism". Gels were fixed in TCA and stained with Coomassie Brilliant Blue R250 . (from Allen) (unpublished).

The Neville and Laemmli systems both employed homogeneous gels as did the earlier CZE system described by Weber and Osborn (64). In attempt to define optimal conditions for such systems, Lanzillo *et al.* (65) utilized a theoretical plate approach to resolution. Unfortunately this study was limited, since no linear or curvi-linear gradients were used to determine if the results obtained in homgeneous systems were analogous to linear gradient systems.

Results of the many reports on molecular weight determination clearly indicate that pore gradient gels offer significant advantages over homogenious gel for such studies. Many of the shortcomings mentioned above can be obviated by utilizing pore gradient systems in conjunction with MZE. In such systems eventually all components will migrate to their pore limit and essentially stop. Hence, unstacked proteins will not cause streaking or be lost due to diffusion, because automatic zone sharpening by pore restriction occurs. Similarly, the non-unit charge of unsaturated lipo- and glycoproteins will not affect their apparent molecular weights. On the surface, it would appear that

MZE buffers lose their advantage in pore gradient systems; however, the voltage gradient provided is highest in the region of the highest molecular size species where it is most useful to force them to their pore limit in a shorter time. Thus, the separation may be achieved in a much shorter run time than with a CZE system. The one precaution is that low molecular size species do not remain stacked on the boundary and consequently not be resolved. Obviously , if the boundary marked by the tracking dye is run off the annodal end of the gel any such stacked proteins will be lost.

Currently linear and concave gradients dominate in this field in conjunction with the Glycine-Chloride MZE buffer system of Laemmli (60).Burghes, *et al.* (66) have modified an exponential gradient tailored to optimize resolution of fibroblast proteins, which demontrates the potential of this technique as illustrated in Fig. 13. Methods of gradient production are given later in the section on gel casting.

From the foregoing it is apparent that for routine determination of molecular weight, or molecular mass (MM), pore gradient MZE-SDS-PAGE is the method of choice for complex and unknown mixtures of macromolecules. Rothe (67) has reported that with gel gradient SDS-PAGE on revaluating the data of Lambin (68) that the accuracy is in the same range for non-denatured and SDS-denatured proteins (8 - 11 %). The analogous behavious of non-denatured and SDS-denatured proteins in gradient gels becomes more evident when time dependent migration distances are determined (67) in linear gradients.

These methods do produce the highest resolution of complex mixtures of macromolecules. Dunn (69) has stated the following: "Resolution in the second dimension can be increased by lengthening the SDS gel. In this respect the SDS gel is analogous to DNA sequencing gels".

While this comment was in reference to two-dimensional studies, the concept for size separation is also important. Anderson and Peterson (70)have shown the value of this technique on 28 cm gels in the identification of gene products. They have demonstrated that up to 150 protein bands can be resolved using the Laemmli technique (60) in 12 - 19 % gradient gels with *B. subtilis* extracts treated with SDS, as shown in Fig. 14.

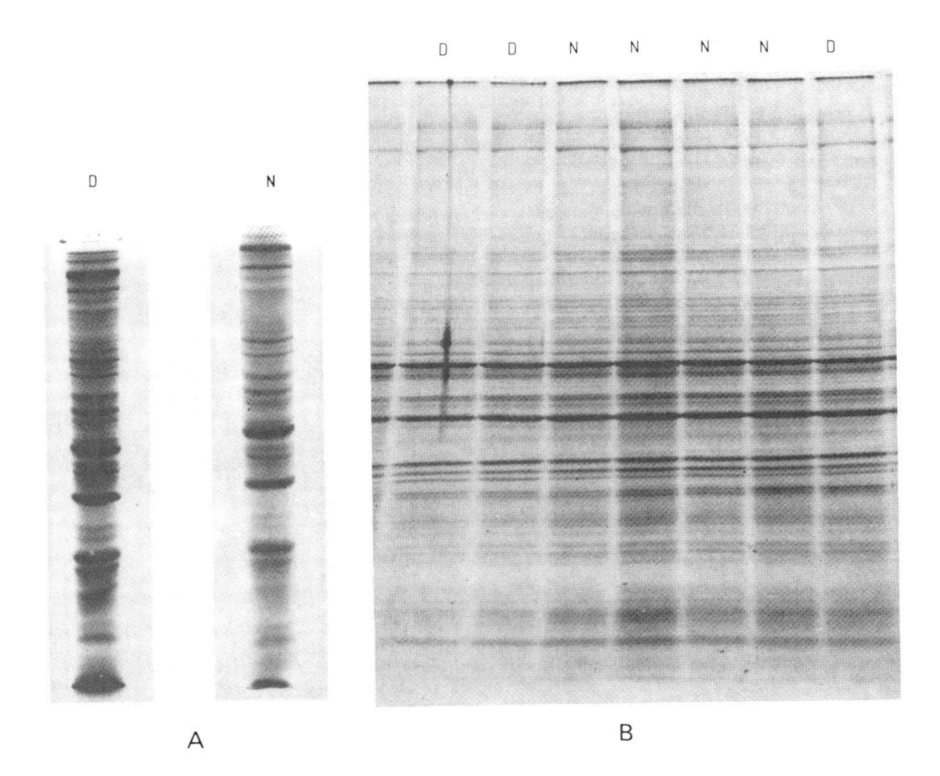

Fig. 13. Comparison of fibroblast proteins separated in a 10 per cent T homogeneous gel in panel A and a 5-20 per cent continuous gradient in panel B. The Laemmli system was used and the gels were stained with Coomassie Brilliant Blue R250. N - Normal and D - Duchennes muscular distrophy. (from Burghes *et al.*) (66).

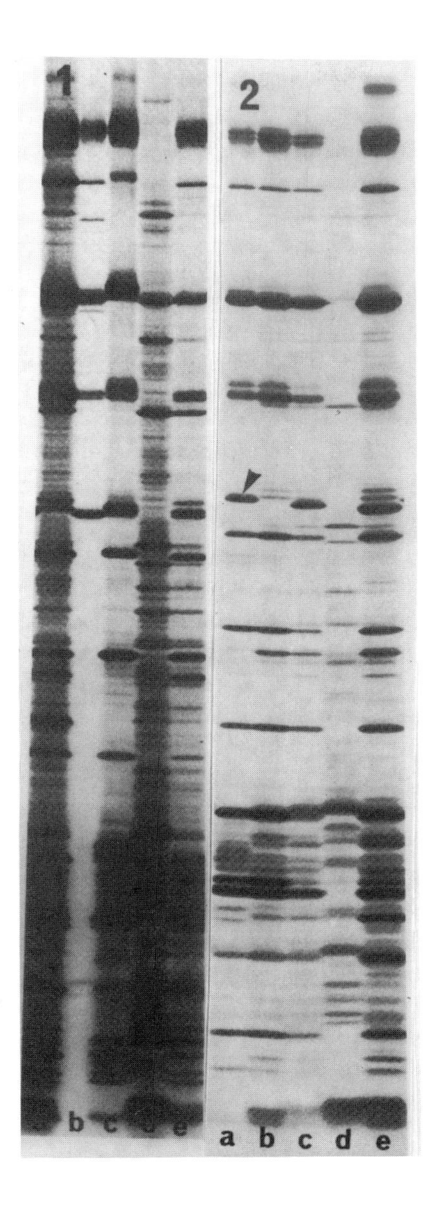

Fig. 14. Autoradiographs of B. subtilis Spo A12 ^{14}C labeled proteins. Cells were infected with 029^{+} phage: a, sus 8.5 (900) infection, b) purified 029^{+} virions, c and e, are 029^{+} cells and d) uninfected control cells. 2. Profiles of pellet and lysates: a) sus 2(628) infection supernatant, b,c and e infections with 900 series phage with a point mutation at 8.5 and d) uninfected control. (from Anderson and Peterson) (70).

2.4. Other Electrochemical Considerations

It is a general rule in electrophoresis that lower ionic strengths of buffers allow higher migration velocities with less heat production, while higher ionic strengths (I > 0.15) lead to lower migration velocities and increased heat generation, although, the zones become sharper (38).

As a close approximation, the mobility is inversely proportional to the square root of the ionic strength. Lowering the ionic strength by a factor of 2 will allow a 1.41 increase in voltage with no increased Joule heating (11). Conversely, reducing the voltage gradient and the ionic strength may lead to similar separation times, less Joule heating and less resolution. Alternatively, reducing the gel thickness will also reduce Joule heating allowing the use of higher ionic strengths, and voltage to improve resolution (see isoelectric focusing Chapter 3).

Similar conditions apply in multiphasic buffer systems. However, the effect of the Kohlrausch regulating function in describing the ionic strength of the trailing ion must be borne in mind. This effect is normally overshadowed by the increasing resistance from the boundary to the origin, which is the major source of Joule heating in moving boundary systems as opposed to continuous buffer systems. In utilizing multiphasic systems in slab gels, where side-by-side comparisons are desired, sufficient ionic strength must be used in the sample so that the ionic strength of the sample and the gel space between samples is equivalent. Otherwise, if the sample has too low an ionic strength, the sample track will pinch together from the Kohlrausch "siphon effect". Precipitation in the sample can also occur from too low an ionic strength, or to inadequate centifugation of tissue extracts; this too often will cause sample pinching. Not only are these affects esthetically unpleasant, but also render the the separation unfit for for quantitative analysis. Moreover, sulfate ions formed from the decomposition of persulfate during polymerization will increase the effective ionic strength, as well as produce an additional boundary in those multiphasic buffer systems employing leading ions other than sulphate. A new procedure for polymerizing acrylamide without the use of catalysts (soon to be released) will obviate this longstanding problem.

2.5. Power Application *vs.* Joule Heat

During every electrophoretic separation, electrical energy is transformed into heat, termed Joule heating, with its potential deleterious affects, such as enzyme denaturation, increased diffusion, *etc.* Joule heating should be kept as low as is consistant with adequate resolution. This may be accomplished by a number of means. Control of electrical power input requirements can be regulated by the electrochemistry of the system (ion species, ionic strength and composition , as outlined in the previous section) or by the method of power delivery to the system in the case of multiphasic buffers. In the case of homogeneous buffers throughout the system (same electrode and gel buffer) at a constant regulated temperature, the power will remain essentially constant throughout the separation and Joule heating may be controlled simply by power input. However, in multiphasic buffer systems- generally used as the preferred method of zone sharpening for increased resolution- the electrical resistance of the gel increases as the boundary migrates through the gel, due to the decrease in conductance. Hence, with constant current application, voltage and consequently Joule heating rise with time. At constant voltage, the current drops with separation time for the same reason, resulting in increased separation times and normally reduced resolution; however, Joule heating is reduced.

The interrelationship of these phenomena are given in the following equations:

$$P = V x I + I^2 x R + V^2/R \text{ (watts)}$$

$$H = V x I / 4.185 \text{ (cal/Sec)}$$

$$R = V/I \text{ (ohms)}$$

Where **P** = power in watts, **I** = current in amperes, **V** = voltage in volts, **R** = resistance in ohms and **H** = Joule heating in cal/sec.

Therefore, it is a safe rule in general, to use constant current in systems with decreasing or constant resistance during the separation and constant voltage in systems where the resistance increases during the separation. In the latter case the separation will slow down impairing resolution due to an increased time for diffusion, particularly in homogeneous pore size gels. It follows that the only way to provide adequate regulation of Joule heating consistant with optimal resolution in multiphasic buffer systems is to use power at a constant level. This may be achieved by using a constant power supply, which automatically senses voltage and amperage and keeps the product of these two parameters constant at any selected value. Alternatively, constant power was first obtained by using pulses of constant power achieved by charging a capacitor under inductance and the discharging it by triggring with a silicon rectifier (19). This method produced constant power pulses, wherein the pulse rate, capacitance and voltage could be independently chosen to produce the power input to the system. In such a configuration the power varies linearly with pulse rate and capacitance, but by the square root of the voltage, as shown in the following equation:

$$\mathbf{P} = 1/2 \times f \times C \times V^2$$

where f = discharge rate in pulses per second, **C** = capacitance in farads and **P** and **V** as above. The advantage of such a system is that a low duty cycle (power off for longer periods than on) can be employed which allows heat dissipation during the off cycle. Also, the Joule heating is calculated from the average voltage; yet in such a system, the peak voltage achieved by charging and discharging the capacitor is 1.7 to 1.8 times higher than the average voltage. This effect is probably of far greater advantage in isoelectric focusing at the end of the separation for zone sharpening, than for multiphasic systems. A consistant 25 % increase in total esterase activity could be demonstrated with this procedure as contrasted to constant current input at the same average power input (39). Unfortunately, to date, such power supplies have only been able to achieve voltage levels of 1800 volts peak.

2.6. Gel Casting

In casting gels it is always an advantage to make up stock solutions that may be used in a set ratio to make various per cent T homogeneous gels. This simplification will eliminate the possibility of errors, particularly when gels are not being cast routinely each day or when different technicians are involved. An eight part ratio of volume of monomer and crosslinker to buffer and catalyst works well and is used in the illustations here of casting homogeneous and simple step gradient gels. Linear and curvilinear gradients are quite another matter and will be handled separately. In Table 7 and 8 below are given the amounts of acrylamide and bis required for different per cent T gels from 3 to 30 per cent.

Table 7. Stock acrylamide solutions 40 and 50 % T

1. Stock 40 % T	3.865 g Recrystallized Acrylamide 0.135 g Bis Dilute to 10 ml with distilled water
2. Stock 50 % Acrylamide	4.830 g Recrystallized acrylamide 0.170 g Bis Dilute to 10 ml with distilled water

To hasten solution of the Bis warm to 37 to 40 °C after addition of the water to the monomer and crosslinking Bis and swirl. Cover before use to prevent evaporation and make fresh daily!

Table 8. Gel preparation table for gels 3 % to 30 % T.

Gel % T	Stock Acryl*	Water	Buffer	Catalyst
3	0.6	2.7	2.0	2.7
4	0.8	2.7	2.0	2.6
5	1.0	2.5	2.0	2.5
6	1.2	2.4	2.0	2.4
7	1.4	2.3	2.0	2.3
8	1.6	2.2	2.0	2.2
9	1.8	2.1	2.0	2.1
10	2.0	2.0	2.0	2.0
11	2.2	1.9	2.0	1.9
12	2.4	1.8	2.0	1.8
13	2.6	1.7	2.0	1.7
14	2.8	1.6	2.0	1.6
15	3.0	1.5	2.0	1.5
16	3.2	1.4	2.0	1.4
17	3.4	1.3	2.0	1.3
18	3.6	1.2	2.0	1.2
19	3.8	1.1	2.0	1.1
20	4.0	1.0	2.0	1.0
21	4.2	0.8	2.0	1.0
22	4.4	0.6	2.0	1.0
24	4.8	0.2	2.0	1.0
26	4.16	0.84	2.0	1.0
28	4.48	0.52	2.0	1.0
30	4.8	0.2	2.0	1.0

* Note change from 40 % stock acrylamide between 24 and 26 % T.

Casting procedures will vary for each individual piece of equipment and the reader is referred to his or her instrument instruction manual for required volumes of gel for each chamber or tube setup. Simple step gradients may be cast by a layering technique, where successively less dense gradients are layered on top of each other with a syringe and narrow gauge canula. These may be layered from the side of the casting chamber and the lighter layers will flow evenly across each preceeding more dense layer. In practice 3 per cent differences are readily layered without difficulty. Visual observation of the gradient formation is facilitated by adding Bromphenol Blue to each alternate solution before layering. Ammonium persulfate addition and degassing should be carried out individually on each suceeding layer just before addition to the cell (see Chapter 1). By layering before polymerization has taken place, the diffusion occurring at the interface of the different density layers allows a smooth transition of the gradient and prevents abrupt per cent T interfaces. The amounts of persulfate in Table 8. are calculated to allow polymerization from the top down, or from the less dense to the more dense in order to prevent heat convection from the exothermic polymerization from distorting the gradient by premature polymerization from the bottom. Casting should be carried out at the temperature that the gel will be run. Examples of a homogeneous gel with glycine - chloride MZE and a simple step gradient gel with borate - sulfate are given in tables 9 and 10. For preparing and selecting the more highly specific MZE buffers for fractionation, the reader is referred to the recent paper of Chrambach and Jovin (38).

Table 9. Buffer stock solutions for Ornstein glycine - chloride system

Separating gel buffer	Spacer gel buffer	Sample buffer
36.6 g Tris 48.0 ml 1NHCl 0.23ml TEMED Dilute to 100 ml with distilled water (pH 8.9)	5.98 g Tris 48.0 ml 1NHCl 0.46 ml TEMED Dilute to 100 ml (pH 6.9)	6.25 ml spacer gel buffer 12.5 g sucrose 11.0 ml water 0.1 ml 0.1 % aqueous Bromphenol Blue

The electrode buffers consists of 6.0 g Tris and 28.8 g glycine diluted to 1000ml (pH 8.3)

Casting of polyacrylamide gels containing AcrylAide crosslinker bound on GelBond -PAG has been reported to produce gels more resistant to fracturing than those containing bisacrylamide. This crosslinker also reduces the possibility of breakage of tube gels during their removal and allows drying of thicker gels witout cracking or peeling, which is particularly useful to produce a permanent record with SDS-PAGE gels.

Although derived from agarose, AcrylAide cross-linker is water soluble at room temperature and does not form a gel when stored in the refrigerator. Instructions are provided below for its use in a typical SDS-PAGE discontinuous slab gel electrophoresis procedure in the Laemmli system. Obviously, Bis or another crosslinking agent could be used in the following formulation.

Table 10. Homogeneous 16.5 % Laemmli gel with AcrylAide crosslinker

1.	Stock solution	Disperse 1 gm of AcrylAide in 50 ml of distilled water at room temperature and stir vigorously using a magnetic stirrer until all of the hydrated particles have dissolved (approximately 1 h). Dissolve 40 gm of acrylamide in the AcrylAide cross-linker solution and adjust the final volume to 100 ml with distilled water. The acrylamide/AcrylAide cross-linker stock solution should then be filtered through Whatman #4 filter paper. The solution is reported, by the manufacturer, to remain stable for a least one month when stored in a dark bottle at 4 ºC.
2.	Separating gel	Use the following formulation for preparing 20 ml of a 16 per cent polyacrylamide, 0.5 per cent AcrylAide separating gel (16.5 per cent T, 3 per cent C equivalent) for SDS - PAGE 3.2 volumes (8.0 ml) of 40 per cent acrylamide, 1 per cent AcrylAide cross-linker stock solution. 1.4 volumes (3.5 ml) of water 2.0 volumes (5.0 ml) of separating buffer (0.75M Tris-HCl and 0.2 per cent SDS (pH 8.8)

Table 10. Cont.

	1.4 volumes (3.5 ml) of ammonium persulfate is added to a final concentration of 0.5 mg/mL and N,N,N^1,N^1-tetramethylethylenediamine (TEMED) at 0.25 µl/ml to initiate polymerization.
	Incorporate a sheet of GelBond -PAG film into your casting apparatus so that the hydrophilic side will be exposed to the acrylamide - AcrylAide cross-linker containing solution. Add the separating gel solution to the casting apparatus and carefully overlay a solution of 0.375M Tris-HCl, 0.1 per cent SDS at pH 8.8.
3.	Following polymerization of the separating gel, discard any remaining fluid from the surface of the gel and insert a comb for forming sample wells. Cast a stacking gel containing 4.85 per cent acrylamide, 0.15 per cent N,N^1-methylene-bisacrylamide in 0.125M Tris-HCl, 0.1 per cent SDS (pH 6.8). Ammonium persulfate is added to a final concentration of 0.5 mg/ml and TEMED at 0.25 µl/mL to initate polymerization.
4.	Boil samples for at least two minutes in pH 6.8 buffer containing 0.0625M Tris-HCL, 2 per cent SDS, 10 per cent glycerol, 0.5 per cent 2-mercaptoethanol (or 0.05 % DTT) and 0.001 per cent phenol red. Apply samples to the sample wells and electrophorese at 25 ma in electrode buffer containing 0.025M **Tris**, 0.192 M glycine and 0.1 per cent SDS at pH 8.3.
5.	Following fixation, staining, and destaining, soak the gel in a 5 per cent glycerol, 5 per cent acetic acid solution for at least two hours. This will enhance flexibility of the dried gel on the plastic support. The gel can then be placed in a forced-air oven and dried to a film at 60 ºC for 2 h.

During fixation or staining, gels can swell considerably. Slab gels bonded to GelBond PAG film will maintain their length and width dimensions while expanding in the thickness dimension. Tube gels expand in length and diameter. This expansion does not, however, effect resolution of the protein bands and can be reversed by gradual dehydration in a methanol/water solution.

2.6.1. Borate - sulfate system in a simple Step gradient

Table 11. Stock solutions for borate - sulfate system

Separating gel soln. 1a	0.75M Tris-sufate 36.3g Tris 62.0 ml 1N H_2SO_4 0.48 ml TEMED Dilute to 200 ml with distilled water (pH 9.0 at 25 ºC)
Well and cap gel Soln. 2a	20 ml of stock 1a 0.36ml TEMED Dilute to 100 ml with distilled water (pH 9.0 at 25ºC)
Sample buffer	6.25ml of stock soln. 2a 12.5 g sucrose 11.0 ml water 0.1 ml 0.1 % aqueous Bromphenol Blue
Electrode buffer	62.88 g Tris 8.74 g Boric acid Dilute to final volume of 8 liters (pH 9.0 at 25ºC) Merthiolate (Thimerosol) may be added at a level of 0.8 g as a preservative

Table 12. Gel casting procedure for a 3 - 6 - 9 -12 % step gradient gel

12 % Layer	2.4 vol 48 % acrylamide stock soln. 1.8 vol water 2.0 vol buffer 1a 1.0 vol persulfate (0.1 %)
9 % Layer	1.8 vol stock acrylamide 2.1 vol water 2.0 vol buffer 1a 2.0 vol persulfate 0.1 vol 0.1 % aqueous Bromphenol Blue

Table 12. Cont.

6 % Layer	1.2 vol. acrylamide - bis stock solution 2.4 vol. water 2.0 vol. buffer 1a 2.4 vol. persufate
3 % Layer	0.6 vol. acrylamide - bis stock solution 2.7 vol. water 2.0 vol. buffer 2a 2.6 vol. persulfate 0.1 vol. 0.1 % aqueous Bromphenol Blue
Well and cap gel	1.2 vol. acrylamide - bis stock solution 2.4 vol. water 2.0 vol. buffer 2a 2.4 vol. 0.07 % persulfate

The casting procedure is performed as follows:

1. Reagents for all of the step gradient are prepared in separate beakers, with the exception that no catalyst is added. The beakers are then covered to prevent evaporation and dust. Exact vol.umes will depend on the apparatus and the depth desired for each gradient step.
2. Persulfate is added to the 12 % layer and the solution degassed and poured into the chamber.
3. Persulfate is next added to the 9 % T gel, followed by degassing and the solution is carefully layered on to the first layer using a syringe with a long 22 gauge canula. Add the solution at one side of the cell so that any gradient distortion at this point will not extend into the adjacent sample tracks. The dye in this solution will provide ready visual monitoring of the layering. As more of the solution is added one can increase the rate of addition.
4. Step 3 is repeated with the remaining solutions. Following addition of the 3 % layer, to the desired height, water layer and allow to polymerize at 4 °C, or at the separating temperature.
5. Following polymerization, normally 45 min, pour off the water layer and add the well gell and sample comb. Here more persulfate can be used to form the wells in 8 - 10 min.
6. Pour off unpolymerized gel, wash the wells with distilled water, pour off and blot the wells dry. Next add the cap gel with 0.07 % persulfate and immediately layer the samples in sucrose and buffer 2a at the well bottom, let polymerize and then the gel is ready to be removed from the casting

device and the electrophoretic separation performed

The sample may be run undenatured, or SDS-denatured and 0.1 % SDS added to the upper buffer tank for separation by size isomerism.

2.6.2. Linear and curvilinear gradients

Linear or curvilinear gradients may be cast with a number of commercially available gradient formers. A schematic of such a device is shown in Fig. 15.

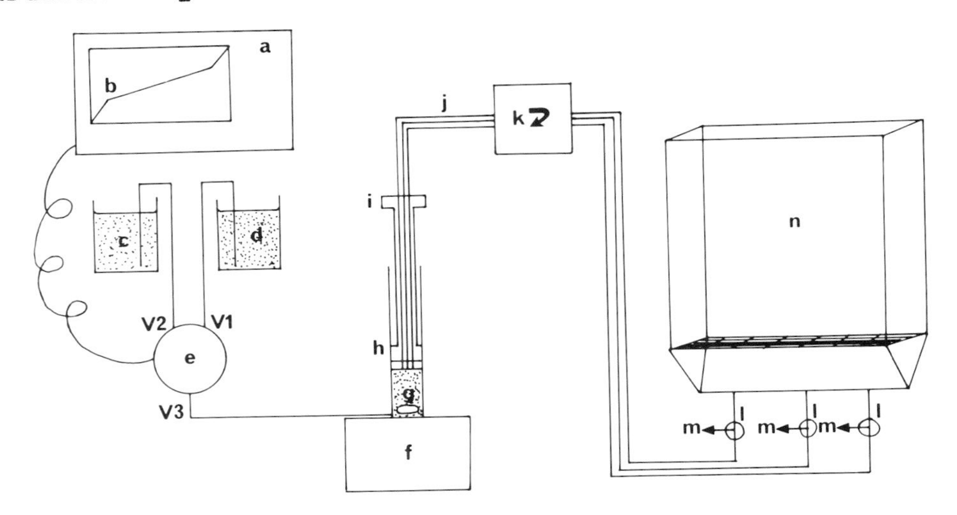

Fig. 15. Gradient forming device for casting various types of gradients using an LKB Ultrograd 11300 (a), fitted with the appropriate gradient template (b). Control is by a three way valve (e) for the dense (c) and light (d) monomer-crosslinker solutions. The solution in the mixing chamber (h) with the plunger (i) is mixed by a magnetic flea (g) and stirrer (f). Fluid flow is controlled by a three channel peristaltic pump (k). All lines (i) are of narrow-bore tubing. The perspex casting tower (n) is fitted with three inlets controlled by three-way taps (l) allowing fluid to be run to waste (m).(from Burghes *et al.*) (69).

More accurately controlled gradients are, perhaps, better cast with the use of a computer controlled gradient maker such as is available from Bio-Rad and shown below in Fig. 16. The low cost and availability of personal computers today would seem to make this the most feasible approach for accurate reproduceability. A number of programs are available and these may also be written in basic to cover all requirements. Altland (72) has produced a number of such programs and found these far more accurate than previous mechanical devices. For those who require a number of gradient gels cast simultaneously, displacement casting of multiple cassettes in a casting tank is required. In this case water is displaced by an increasingly dense solution of acrylamide, buffer and catalyst to a desired gradient height. Here the gel is automatically waterlayered to promote polymerization and to produce an optically flat surface for the even entry of sample. Also buffer may be used in lieu of water.

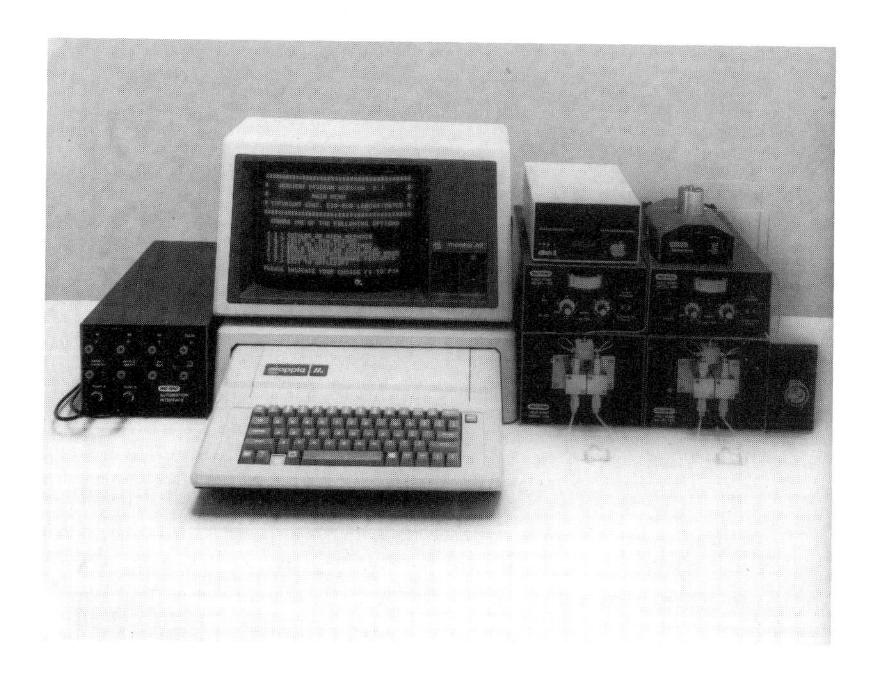

Fig. 16.. Computer operated gradient forming device (from Bio-Rad)

2.7. References

1. Chrambach, A., Jovin, T. M., Svendsen, P. J., and Rodbard, D. : in Catsimpoolas, Editor, *Methods in Protein Separation 2,* Plenum Publishing Co., N. Y., p.27 (1976).

2. Rodbard, D. and Chrambach, A. : in Allen, R. C. and Maurer, H. R. Editors, *Electrophoresis and Isoelectric Focusing in Polyacrylamide Gel* deGruyter, Berlin p. 28 (1974).

3. Morris, C. J. O. R., and Morris, P. : *Biochem. J. 124,* 517 (1971).

4. Hjerten, S., Jerstedt, S., and Tiselius, A. : *Anal. Biochem. 11,* 219 (1965).

6. Ornstein, L. and Davis, B. J. : *Disc Electrophoresis : Parts I and II,* Distillation Industries, Rochester, N. Y. (1962).

7. Slater, G. G. : *Federation Proc. 24,* 225 (1965).

8. Slater, G. G. : *Anal. Biochem. 24,* 215 (1968).

9. Margolis, J., and Kendrick, K. G. : *Anal. Biochem. 25,* 347 (1968).

10. Poulik. M. D. : *Nature 180,* 1477 (1957).

11. Ornstein, L. : *Ann. N. Y. Acad. Sci. 121,* 321 (1964).

12. Kohlrausch, F. : *Ann. d. Phys. u. Chem. 62,* 209 (1897).

13. Kendall,J., Jette, E.R., and West. W. : *J. Am. Chem. Soc. 48,* 3114 (1926).

14. Kendall, J. and Clarke, B. L. : *Proc. Nat. Acad. Sci. 11,* 393 (1925).

15. Kendall, J. and West, W. : *J. Am. Chem. Soc. 48,* 2619 (1926).

16. Kendall, J. : *Science 67,* 163 (1928).

17. Allen, R. C., Popp, R. A., and Moore, D. J. : *J. Histochem. Cytochem. 13,* 249, (1965).

18. *Disc Electrophoresis Newsletter 9,* Canalco Inc., Rockville, MD p.8 (1967).

19. Allen, R. C., Moore, D. J., and Dilworth, R. H. : *J. Histochem. Cytochem. 17,* 189 (1969).

20. Allen, R. C., and Moore, D, J. : *US Patent No. 3,620,947* (1971).

21. Allen, R. C. : in Allen, R. C. and Maurer, H. R., Editors,*Electrophoresis and Isoelectric Focusing in Polyacrylamide Gel* de Gruter, Berlin, p. 105 (1974).

22. Maurer, H. R. and Allen, R. C. : *Z. Klin. Chem. Klin. Biochem. 10,* 220 (1972).

23. Jovin, T. M., Dante, M. L., and Chrambach, A. : *Multiphasic Buffer Systems Output,* National Technical Information Service, Springfield, Va. 22151, PB No. 196085-196091, 203016 (1970).

24. Jovin, T. M. : *Biochemistry 12,* 871 (1973).

25. Everearts, F. M., Beckers, J. L., and Verheggen, T. P. E. M. : *Isotachophoresis* , Elsevier, Amsterdam (1973).

26. Schafer-Nielsen, C. and Svendsen, P. J. : *Anal. Biochem. 114,* 244 (1981).

27. Svensson, H. : *Ark. Chemi Mineral Geol. 17A,* 1 (1943).

28. Svensson, H. : *Sci. Tools 3,* 30 (1956).

29. Hjelmaland, L. M. and Chrambach, A. : *Electrophoresis 3,* 9 (1981).

30. Martin A. J. P. and Everaerts, F. M. : *Proc. Roy. Soc. Ser. A. 316,* 493 (1970).

31. Hjerten, S., Oferstedt, L.-G., and Johansson, G. : *J. Chromatogr. 194,* 1 (1980).

32. Hjelmaland, L. M. and Chrambach, A. : *Electrophoresis 4,* 20 (1983).

33. Buzas, Z., Hjelmeland, L. G. and Chrambach, A. : *Electrophoresis 4,* 27 (1983).

38. Chrambach, A. and Jovin, M. : *Electrophoresis 4,* 190 (1983).

39. Rodbard, D, and Chrambach, A. : *Proc. Natl. Acad. Sci. USA 65,* 970 (1970).

40. Chrambach, A.: *Mol. Cell Biochem. 29* , 23 (1980).

41. Hedrick, J. L., and Smith, A. J. : *Arch. Biochem. Biophys. 126,* 155 (1968).

42. Gonenne, A. and Lebowitz, J. : *Anal. Biochem. 64,* 414 (1975).

43. Rothe, G. M., and Purkhandaba, H. : *Electrophoresis 3,* 33 (1982).

44. Rothe, G. M. and Purkhandaba, H. : *Electrophoresis 3,* 43 (1982).

45. Rodbard, D., and Chrambach, A. : *Anal. Biochem. 40,* 95 (1971).

46. Felgenhauer, K. : *Hoppe-Seyler's Z. Physiol. Chem. 355* 1281 (1974).

47. Kopperschläger, G., Diezel, W., Bierwagen, B., and Hofmann, E. : *FEBS Letters 5,* 221 (1969).

48. Andersson, L. O., Borg, H., and Mikaelsson, M. : *FEBS Letters 20,* 199 (1972).

49. Lasky, M., : in Catsimpoolas, N. (Editor), *Electrophoresis '78,* Elsevier, North Holland, New York, p. 195 (1978).

50. Slater, G. G. : *Anal. Chem. 41,* 1039 (1969).

51. Rodbard, D., Kapadia, G., and Chrambach, A. : *Anal. Biochem. 40,* 95 (1971).

52. Lambin, P., and Fine, J. M. : *Anal. Biochem. 98,* 160 (1979).

53. Shapiro, A. L., Vinuela, E., and Maizel, J. V. : *Biochem. Biophys. Res. Commun. 28,* 815 (1967).

54. Maizel, J. V. ; in Maramorosch, F, and Koprowski, H.,(Editors), *Methods of Virology 5,* Academic Press, New York, p. 179 (1971).

55. Reeder, D. J., Davidson, A. E., Parris, R. M., Reubert, R. E., and Chesler, S. N. : *Abst. Boston '83,* (1983).

56. Reeder, D. J. : Personal communication.

57. An der Lan, B., Cowell, J. L., Burstyn, D. G., Manclark, C. E., and Chrambach, A. : *Arch. Biochem. Biophys.*, In Press (1984).

58. Rodbard, D. : in Catsimpoolas, N. Editor, *Methods of Protein Separation 2*, Plenum Press, New York, p. 145 (1976).

59. Shah, A. A. : *Electrophoresis 5*, in Press.

60. Laemmeli, U. K. : *Nature 227*, 680 (1970).

61. Wykoff, M., Rodbard, D., and Chrambach, A. : *Anal. Biochem. 78*, 459 (1977).

62. Neville, D. M. : *J. Biol. Chem. 246*, 6328 (1971).

63. Chrambach, A. : Personal Communication.

64. Weber, K. and Osborn, M. J. : *J. Biol. Chem. 244*, 4406 (1969).

65. Lanzillo, J. J., Stevens, J. and Fanburg, B. L. : *Electrophoresis 1*, 180 (1980).

66. Burghes, A. H. M., Dunn, M. J., Statham, H. E. and Dubowitz, V. : *Electrophoresis 3*, 177 (1982).

67. Rothe, G. M. : *Electrophoresis 3*, 255 (1982).

68. Lambin, P. : *Anal. Biochem. 85*, 160 (1978).

69. Dunn, M. J. and Burghes, A. H. M. : *Electrophoresis 4*, 97 (1983).

70. Anderson, D. and Peterson, C. : in Allen, R. C. and Arnaud, P. Editors, *Electrophoresis '81* de Gruyter, Berlin, p. 41 (1981).

71. Altland, K. : Personal communication.

3. ISOELECTRIC FOCUSING

3.1. Introduction

Since the last edition of this book, isoelectric focusing in polyacrylamide gel (PAGIF) and in charge-free agarose (AGIF) has literally exploded as a separation technique for amphoteric macromolecules. The historical development of this method has been described in detail by Righetti and Drysdale (1), Righetti (2) and in the book *Electrokinetic Separation Methods* (3), all of which provide excellent general reading. In addition, a series of recent reviews on such areas as preparative isoelectric focusing (4), biochemical and clinical applications (5), general applications (6), methodological aspects in comparison with electrophoresis (7) and recent developments (8) have appeared. In addition the International Congresses of Electrophoresis have published proceedings dealing with new developments and applications. The latest of these were held in Charleston in 1981 (9), Athens in 1982 (10) and Tokyo in 1983 (11).

3.2. Operation

All ampholytes, including proteins, have an isoelectric point (pI) which cause them to be negatively charged in solutions above their isoelectric point and positively charged in solutions where the pH is below their isoelectric point. Since the surface charge determines the direction of migration in an electrical field, it follows that at pH levels above the pI of an ampholyte it will migrate toward the anode, and conversely, toward the cathode in solutions where the pH is below the isoelectric point. While at its isoelectric point, it will achieve a steady state of zero migration.

The development of practical laboratory methods for isoelectric focusing has added a new dimension to electrophoretic techniques in polyacrylamide and charge-free agarose support media for the separation of proteins and other amphoteric substances. Isoelectric focusing, which has been recently demonstrated to be a special case of the moving boundary theory (12) is thus, basically a method for separation of substances according to their isoelectric points in a stable pH gradient whose range may be preselected. For all practical purposes, this may be considered an equilibrium process where zone width is at a minimum at the completion of separation in contrast to zone and moving boundary electrophoresis where the zone width increases with run time due to diffusion. This method offers the advantage of being able to separate amphoteric substances based on their isoelectric points, which is a more valuable characteristic than their mobiliy at a certain pH. Further, the technique may be employed for both analytical and preparative purposes on a variety of supporting media, such as sucrose density gradients, acrylamide, granular gel beds and charge-free agarose gels. The procedure has been refined to the point where one can reproduceably separate substances differing by less than 0.005 pH unit when narrow pH range ampholytes at high voltage gradients are used (13) and to 0.001 pH unit on the newer immobilized pH gradients (14).

3.3. Background

Kolin (15) described the separation of proteins by electrical transport in a pH gradient generated in a sucrose density gradient contained in a Tiselius - like apparatus. The material to be separated was placed at the interface of an acidic and basic buffer which were allowed to diffuse against one another, while at the same time being subjected to an electromotive force. The separation in these experiments was contingent on several interacting factors: a pH gradient, a density gradient, a conductivity gradient and a temperature gradient.

> Kolin in his first paper on separation by this method stated, "The present process is distinct from the conventional electrophoretic separation in that the latter sorts ions according to difference in mobility, whereas, the former separated them according to their isoelectric pH values. It remains to be seen whether or not a general method can be developed for obtaining an isoelectric spectrum of a protein mixture whose components could be identified by their isoelectric points. " He created the term *isoelectric spectra* and ion *focusing* from these early studies.

Such pH gradients were unsuitable due to the rapid migration of the buffers during electrophoresis and the components separated by this procedure were not readily recoverable. However, this work defined two basic points required for optimal performance of this procedure: 1, the ampholytes should have a high buffering capacity and 2, there must be stabilization against convective mixing.

It remained for Svensson (now Rilbe) to introduce the concept of natural pH gradients which established the theoretical basis for isoelectric focusing in its presently used form. The most critical of these concepts states that a natural pH gradient as developed by the electrical current applied to the system is positive throughout the gradient as the pH increases steadily and monotonically from anode to cathode. Therefore, reversal of the pH gradient at any point between the anode and cathode is incompatible with the steady state (16). This advanced the concept of developing a *natural* pH gradient by electrolysis of amphoteric molecules and predicted that stable pH gradients would be obtained by isoelectric stacking of a large series of carrier ampholytes arranged under an electromotive force in order of increasing pI from anode to cathode (17,18). Thus, an important requirement for a good carrier ampholyte is that it has good conductivity at its isoelectric point as well as good buffering capacity. Svensson's theories and specifications of suitable ampholytes led to an extensive search for commercial development.

However, it was his former student Vesterberg (19) who synthesized the first mixture of a large number of homologs and isomers of aliphatic polyaminocarboxylic acid with different pH's and isoelectric points closely spaced between pH 3 to 10. This material was commercialized by LKB under the trade name Ampholine. Several years later Pogacar and Jarecki (20) and Grubhofer and Borjia (21) coupled tetraethylenepentamine and pentaethylenehexamine, propansultone, vinylsulfonate or chloromethylphosphonate which produced sulfonic ampholytes, but which were only suitable for focusing in pH ranges 2 to 3.5 and 6 to 9.5, having little buffering capacity in the pH region of 3.5 to 6.0. These, mixed with aliphatic polycarboxylic acids to fill the pH gap, have been produced and made commercially available by Serva Fine Biochemicals, under the name Servalyte. Until recently these were also distributed by Bio-Rad under the name Biolyte. However this company has now a new material produced by them.

The third commercially available ampholyte was developed by Pharmacia under the trade name Pharmalyte. This is produced in each of five narrow range intervals by a direct synthesis. The principle of the procedure is the condensation of glycine, glycylglycine and amines with selected pKs with epichlorohydrin (22). The choice of amines used in the synthesis allows the production of ampholytes with defined properties that will generate an interval over a given pH range. Mixture of the five narrow range materials provides a pH range of 3 to 10. These ampholytes have an excellent conductivity profile across the entire pH range as evidenced by no local hot spots even at voltage gradients of 500 V/cm and above. This is difficult to do with the previously mentioned ampholytes without additional ampholyte or a mixture of different types. The manufacturers have calculated 1000 ampholyte species per pH unit which should result in more stable pH gradients (22).

In the last year two new ampholytes have been been introduced, one by Separation Sciences and the other by Isolab. These have to date been much less characterized than those mentioned previously and their advantages or disadvantages remain to be elucidated. Both appear to be based on materials containing polypeptide bonds.

3.3.1. New synthetic carrier ampholyte (SCAM) approaches

A new synthetic carrier ampholyte (SCAM) has been described which has a number of advantageous characteristics. This material developed by Bjellquist *et al.* (14) has been designated Immobiline™ in analogy to Ampholine™. It is a set of buffering monomers, a series of acylamide derivatives with the general formula: CH_2= CH-C-N-R, where R contains either a carboxyl or a tertiary amino acid. Known pH values are mixed to generate a linear pH gradient in the desired pH interval. The pH gradient gels are cast in the same way as pore gradient gels, but instead of varying the acrylamide content, the light and heavy solutions are made to different pH values with the immobilizable buffers.

This material is reported to show a number of advantages, of which the most important are: 1, cathodal drift is abolished; 2, very narrow pH ranges may be established giving higher resolution; 3, it allows high load capacities and high salt concentrations in the sample; 4, it has a uniform conductivity and buffering capacity at low concentrations allowing high voltage gradients; 5, it represents a mileau of known and controlled ionic strength some 100 fold less than conventional synthetic carrier ampholytes (SCAMS). While this new development would appear to offer great advantages, the requirement to form a gradient for each separation is somewhat tedious and of doubtful reproducibility between laboratories. Also, at the low ionic strengths presently employed, it requires special effort to measure the pH gradient. Only the test of time will indicate the broadspread usefulness of this most recent development.

3.3.2. Simple buffer ampholytes

Another approach is isoelectric focusing in mixtures of either amphoteric or non-amphoteric simple buffers. This was first demonstrated in two buffers by Kolin in 1954 (23) and later with more buffers (24). However, none of these, or several additional such systems (25,26) were practical in the application to isoelectric focusing, in view of their relative non-linearity and limited pH range. By 1980 two such systems had been developed, one over a range of 3 to 7, as a analytical procedure (27) and the other over a pH range of 4 to 8 on Sephadex gels as a preparative procedure (28). Prior to this time no simple buffer pH gradients covering the entire normal pH range of 3 to 10 had been developed successfully. Cuono and Chapo in 1982 (29) described gel slab PAGIF in a natural pH buffer gradient of 3 to 10 generated by a 47-component buffer mixture. The gradient was stable for at least 18 hours at 100 V/cm with attainment of a steady state at one hour. Buffer constituents are at a final concentration of 10 to 20 mM and have a conductivity approximately 1/10 that of any of the synthetic carrier ampholytes described above. They thereby permit a high degree of resolution, and indeed, allow definition of the heterogeneity of acid α_1 - glycoprotein on a 3 to 10 pH gradient, which is not achieved normally, even on narrower gradients with conventional SCAMS. A distinct advantage of buffer gradients is the lack of lot to lot variation encounterd with conventional SCAMS. An additional advantage lies in the fact that these low molecular weight amphoteric and non-amphoteric substances will not complex with pseudo-proteins (30) nor presumably with proteins which may lead to some of the controvercial results reported by SCAMS including the newer Immobilines (31). Like the Immobiline, cathodal drift has been effectively eliminated as evidenced by pH gradient stability for extended periods of up to 72 hours. However, since there are a number of non-amphoteric subtances in this system, one must to a great degree rely on diffusion for some of the components to focus. Thus, high voltage gradient ultrathin-layer separations may require longer run times than with conventional SCAMS. The future of these systems appears to be of great interest and the method could provide a new level of capability in the area of isoelectric focusing.

3.4. Factors Affecting Resolution in Isoelectric Focusing

The resolution potential of isoelectric focusing is contingent on a number of interdependent factors, such as conductivity, a uniform conductance across the entire gradient, length of the gel, voltage gradient, and volume of the gel, all of which effect Joule heat production and thus the conditions that may be used in a given separation.

3.4.1. Length of the gel

Resolution is proportional to the length for a given field strength (32).

$$\text{Resolution} = \sqrt{\text{Gel Length}}$$

If one doubles the length of the gel at a given voltage gradient, any two adjacent zones will be twice as far apart, but due to the flatter pH gradient, the width of the zones are increased in relation to the equation of Rilbe (33) for the width of a focused zone.

$$X_1 = \pm\sqrt{\frac{D}{E\,\frac{du}{d(pH)}\,\frac{d(pH)}{d^2}}}$$

Where D = diffusion coefficient and E = field strength

The same relationship is shown by Rilbe's equation for the minimum difference in pI for two zones to be resolved.

$$(pI) = 3\sqrt{\frac{D\ d(pH)/dx}{E\ (-du/d(pH))}}$$

Where [d(pH)/dx] is the pH gradient and du/d(pH) the mobility slope.

This is essentially the same conclusion deduced by Giddings and Dahlgren (34).

3.4.2. Resolving power

$$R_s = \sqrt{\frac{1}{\text{pH Gradient}}}$$

3.4.3. Voltage gradient

Resolving power is equal to the square root of the voltage gradient.

$$R_s = \pm\sqrt{DE^{-1} \ idu/d(pH)^{-1} \ d(pH)/(dx)^{-1}}$$

Giddings and Dahlgren (34) state that the resolving power R_s in isoelectric focusing is as follows:

$$R_s = \frac{pH}{4}\sqrt{\frac{-FE \ dg/d(pH)}{RT \ d(pH)/dx}}$$

Where **F** = the faraday (96,000 coulombs), R- the gas constant and T = the absolute temperature

3.4.4. Peak capacity

These authors have defined the peak capacity "n" as the maximum number of components resolvable by a given technique under specified conditions and in isoelectric focusing is given by the following equation:

$$n = \sqrt{\frac{-FE \ dg/d(pH) \cdot d(pH)/dx \ L^2}{L \ 6 \ R \ T}}$$

Where L = length

This equation shows that peak capacity increases in proportion to the total path length and is a function of the square root of both the slope of the pH gradient and the amphoteric macromolecules mobility slope. Assuming a uniform gradient model, or a constant value of [d(pH)/dx], the pH gradient and du/d(pH), the mobility slope.

$$n = \pm\sqrt{\frac{-FE\ dg/d(pH) \cdot d(pH)/dx\ L^2}{16\ RT}}$$

Therefore, peak capacity or theoretical resolution, is directly proportional to the square root of the electric field, path length and total pH increment. These parameters may be varied simultaneously or independently according to experimental desire.

3.5. Practical Condsiderations

Over fifteen years of experience with isoelectric focusing in gels has shown, not withstanding the extreme usefulness of the method, that synthetic carrier ampholytes (SCAMS) have certain inherent characteristics that must be recognized by the user. The first of these is that the patterns produced by ampholytes produced by different manufacturers will not produce precisely the same patterns on a given sample. This is illustrated in Fig. 1., where three different ampholytes were used to separate Pharmacia's Pi standard.

The second problem most commonly described for isoelectric focusing with SCAMS is that of"Cathodic Drift" or the "Plateau Phenomenon" (35-37) . This is the continuous slow change of the pH gradient with time. This phenonenon has been reviewed recently in depth by Righetti (38).

As a general rule pH gradients in polyacrylamide gels are stable, *i.e.*, no cathodal drift up to four hours in gels with or without 20 per cent sucrose and up to eight hours in the presence of stabilizing urea at voltage gradients up to 80 V/cm (39). In ultrathin-layer gels up to 375μ thick, at voltage gradients as great as 460 V/cm with four per cent ampholyte, the gradient remains stable at least for 30-40 minutes after completion of gradient formation as evidenced by increasing resolution at both the anodic and cathodic ends as shown in Fig. 2. (13).

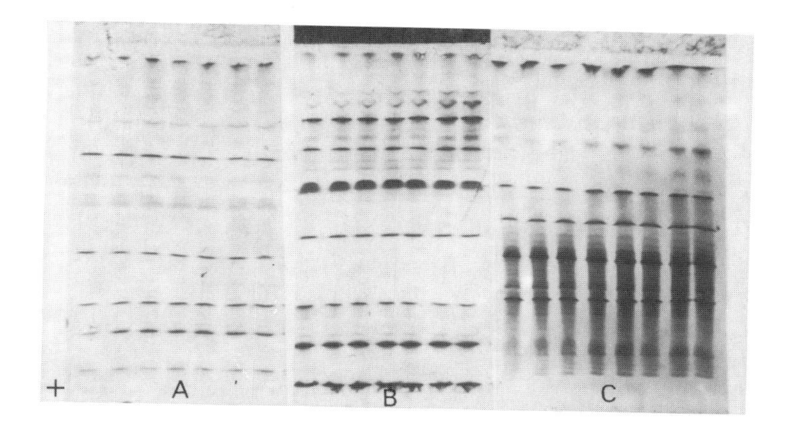

Fig. 1. Separation of Pharmacia pI standard on 375µ four per cent ampholyte gels using pH 3.5 - 10 Biolyte, A; 3.5 - 10 Ampholine, B and 3 - 10 Pharmalyte, C. Separations were carried out at a final sharpening voltage of 350V/cm and proteins were stained with diammine silver. (from Allen *et al.*(41)

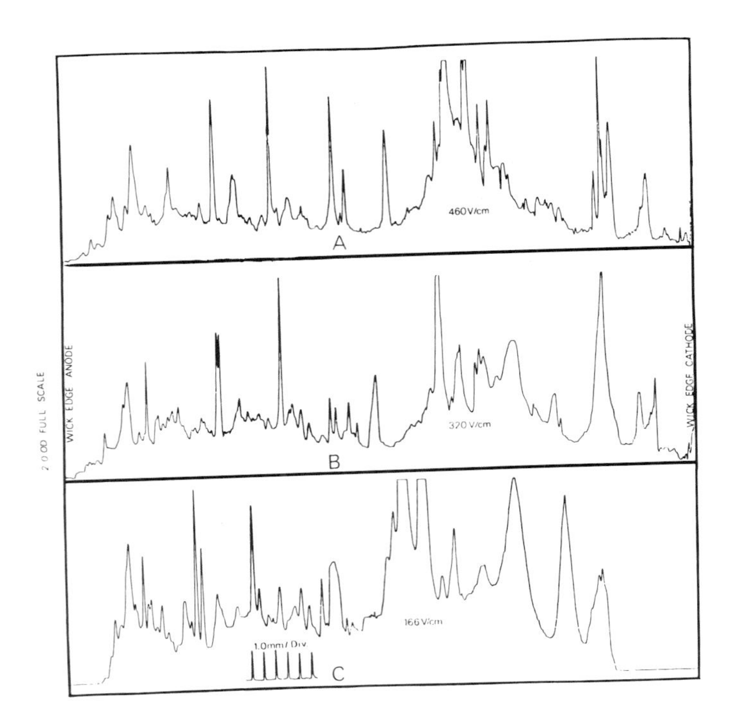

Fig. 2. Densitometric traces of separations of 1.0 µl of human granulocyte extracts with increasing final voltage gradients applied after the gradient was complete. A, 460 V/cm; B, 320 V/cm and C, 166 V/cm. (from Allen) (42)

3.6. Practical Application of Theory

While it is clear that resolution is proportional to the square root of the voltage gradient and the length of the gel, realizing optimal resolution in practice requires a number of considerations with presently existing commercial SCAMS and equipment. Rather than try to approach each parameter individually, since they are mutually interactive; it is better for clarity to discuss the total system.

In any system a fundamental problem with increasing voltage gradient is in the rapid rise in Joule heat which will, at a given point in a given system, literally burn the gel at the point of lowest conductance (or highest region of resistance). Thus, adequate cooling is a must, if the advantage of higher voltage gradients, increased ampholyte concentrations and heat transfer are to be exploited. This consideration has led to the use of flat slab gels with their better inherent heat dissipation characteristics in comparison with gel rods. Early commercially available system utilized liquid cooled plates with gels 2mm thick. It was immediately recognized that thinner flat slabs provided better heat dissipation. By 1978, Görg, *et al.* (40) , had developed ultrathin-layer gels of 125 to 250μ thickness backed on cellophane which provided improved heat dissipation with the then available cooling systems. Radola (41) utilized 20-50μ gels for even better cooling and higher voltage gradients. However, it was apparent that heat dissipation was still limiting in systems employing circulating liquid coolants and plastic or glass cooling plates. Allen, *et al.* (42) described the use of Peltier cooling in conjunction with beryllium oxide ceramic cooling plates. The latter are some 256 times more efficient than glass and have a high dielectric constant. Thus, the cooling plate with the gel cast directly on it, could rapidly be brought to as low as -20 ºC to dissipate Joule heat more effectively. This system was adapted to ultrathin-layer gels (13) and practical voltage gradients of up to 700 V/cm are possible on gels backed in thin polyester sheets using four per cent ampholyte concentrations.

It should be noted here, in the practical separation on flat slabs (either horizontal or vertical) and in gel rods, initial voltage gradients applied are usually on the order 12.5 V/cm on gel rods and 25-30 V/cm in ultrathin gel slabs . This is required since conductance of the system of any pH gradient of SCAMS is at a maximum at the start of the separation with all components being distributed equally throughout the system. High voltage gradients initially produce excessive Joule heating with disasterous results.

The advantageous effects of Peltier cooling and better heat conductivity away from the gel have led to the recent introduction of two such devices for horizontal slab gels. One from MRA Corporation utilizing beryllium oxide plates and one from Hoeffer utilizing a copper- alumina sandwich for the cooling plates. The latter cooling plate is less expensive, but some six times less as effective a heat conductor as is the beryllium oxide. An example of high voltage, ultrathin-layer, isoelectric focusing on the MRA apparatus is shown in Fig. 3.

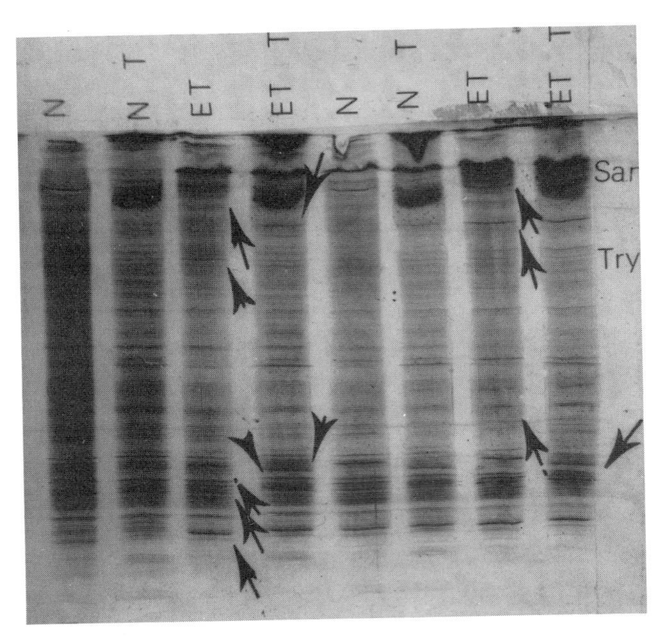

Fig. 3. Macrophage extracts from mice with 1.0µl samples from mice treated with endotoxin. Trypsin was placed 1cm below the samples and allowed to cross migrate, deleting trypsin binding proteins and forming new complexes. The gel was 125µ thick with 4 % Servalyte 3-7 and the maximum final voltage was 600 V/cm and total separation time was 29 min. (from Allen)

3.6.1. Gel thickness effect on resolution

In Fig. 4. are shown sections of three gels prepared simultaneously with thicknesses of 125µ, 250µ and 375µ, respectively. These gels were loaded with 1.0µl of identical samples and focused under maximum end cooling conditions on an MRA "Cold Focus" apparatus utilizing an EC 600, 4000 volt constant power supply. Run conditions were identical, with prefocusing for 10 minutes at 220V at 1 watt. The samples were loaded and run for five minutes at 280V, 1 watt for an additional five minutes at 2.5 watts, five minutes at 5 watts, four minutes at 10 watts and three minutes at a setting of 2800V or 500 V/cm. In the case of the 125µ thick gel total final power was 13 watts and the voltage remained constant. With the 250µ thick gel the voltage dropped to 2300 at a wattage of 33 watts, while the 375µ thick gel dropped to 1900V at a final 45 watts of power. Thus, with the thicker gels the cooling capacity of gels bonded to glass plates with silane, was exceeded.

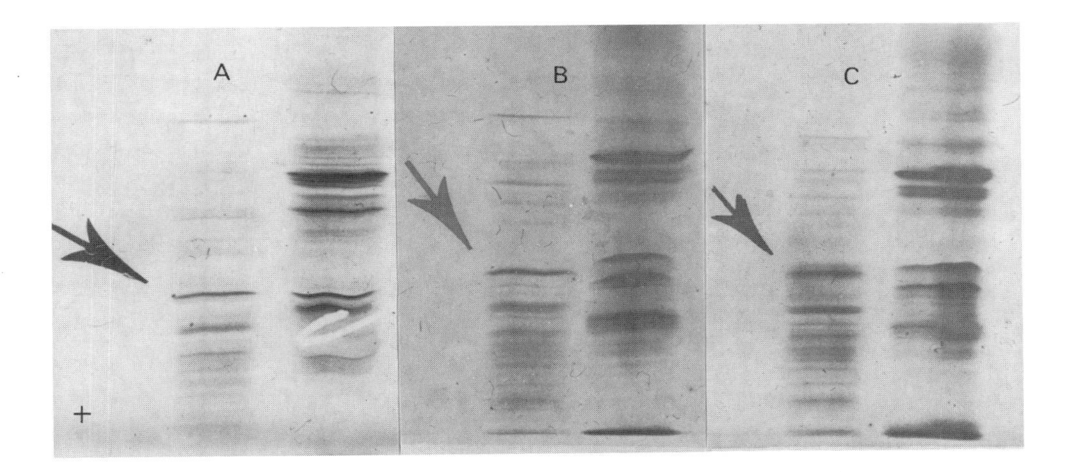

Fig. 4. Limiting effect of gel thickness on heat dissipation and avaiailable voltage gradient: A,125µ; B, 250µ; C, 375µ thick gels.

The effect on resolution is even more marked with the band width (marked with the large arrows) increasing from 100µ to 200µ to 400µ as the gel thickness increased. Since these effects could have resulted partially, or in total, from protein leaching during TCA fixation, the silver stained gels were subjected to independent quantitative analysis, (kindly performed by Professor V. Neuhoff, Göttingen), who determined with automated densitometry, that quantitative reproducibility of these three bands was plus or minus two per cent. Thus, the effect was one of voltage gradient and not due to protein loss during fixation or staining.

3.6.2. Volt hours, voltage gradient and resolution

In the previous section the effect of Joule heating and dropping voltage was illustrated. That the differences in resolution are not due to a difference in the total volt/hour input may be illustrated in Fig. 5. Gels were prepared from a single mix at one time, but run at different voltage gradients for the same number of volt hours with a standard sample of Rohament P extract. Volt/hours were determined using a Pharmacia 3000 power supply with a volt/hour integrator. The concept of volt/hours has recently come into wide use in isoelectric focusing, particularly as a standard for reproducibility of the gel rod focusing commonly utilized in dissociating-denaturing PAGIF-SDS PAGE 2-D procedures. However, these values are only meaningful when the systems on which they are being employed are totally defined. Temperature differences will affect resistance and voltage according to E = IR on systems run at constant power. Voltage will drop if temperature rises increasing current. Total volume (3.6.1) also affects I and thus E. The use of $V/h/cm^3$ allows a more meaningful comparison between different systems.

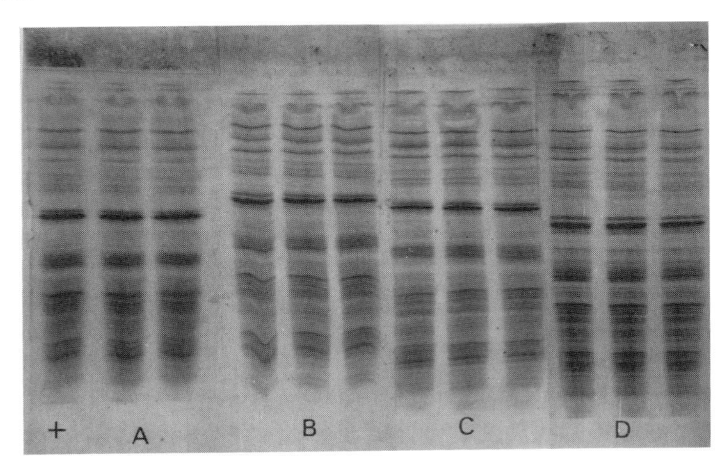

Fig. 5. PAGIF of 1.0 µl of a solution of Rohament P. Separations were carried out on a pH 3 - 7 Servalyte gradient with a gel thickness of 125µ for 450 Vh with maximum voltage gradients of A, 50 V/cm; B, 125V/cm; C,250 V/cm and D, 500V/cm. Separation times were 104 min., 52 min., 39 min. and 29 min. respectively. (from Allen and Arnaud) (43)

3.6.3. Increasing gel length

The dependency of resolution on voltage gradient for a given gel length precludes the obvious choice of simply increasing gel length to increase resolution beyond certain limits. Voltage gradients of 400 V/cm and above can achieve high resolution in acrylamide systems. At present, the authors are unaware of commercially available power supplies of over 6000V capacity, thus, limiting the length of the gel to 8-10 cm. Increased length effects cannot be advantageously employed without development of more powerful regulated power supplies.

However, the choice of narrow range pH gradients can effectively serve the same purpose as longer gels. For example, 2 pH unit gels could give the same theoretical equivalent resolution, at 500V/cm on four 8 cm gels over a pH range of 2 to 10, which could be expected with a 32 cm gel. But, which would also require a 16,000 volt power supply in the latter to achieve 500 V/cm. The use of narrow pH domains, therefore, is a practical alternative to longer gels at lower voltage gradients. In practice, gels of three pH units require relatively short run times to reach full equilibrium conditions, usually less than one hour in ultrathin-layer gels. While additional setup time may be required, the shorter separation times and increased resolution appear to offer an attractive alternative to longer, broad pH range gels, run at lower voltage gradients; in terms of both increased resolution and a lack of distorting cathodal drift. Present methods utilizing isoelectric focusing as the first dimension for 2-D separations rely, in the main, on 16 cm rod gels which require much longer separation times at voltage gradients limited to 40 to 50 V/cm. This problem has been solved (see Chapter 4), but as yet remains to be accepted by many investigators who are expert in the O'Farrell (44) or Iso-Dalt based systems (45,46).

3.6.4. Poor resolution or the "Wavy Band" phenomenon in horizontal PAGIF

Uneveness of the bands across the gel may be caused by a number of factors: The most important of these are: (1) a high salt concentration in the sample with a low salt concentration between sample tracks; (2) uneven electrode pressure or too thin electrode wires; (3) too rapid establishment of the gradient; (4) bands migrating on an angle vertical to the plane of the gel due to a temperature gradient; (5) poor choice of ampholyte or anode buffer; and (6) impurities in the separation media.

Salt concentration effects may be minimized by either dialysing the sample or by applying less sample to the gel. With the newer high resolution techniques, with dilution of sample and silver staining, this problem is effectively obviated. Also the new Immobilines are far less sensitive to salt effects than conventional SCAMS Samples of the same type such as serum can be applied on paper tabs almost touching each other, with a reduction of space between samples and resultant reduction in any salt effects.

A major problem with early commercial systems was that the diameter of the electrode wire was often too small which contributed to poor or uneven contact leading to resolution problems. Increasing wire diameter to 0.030 inch, or the use of the present ribbon electrodes, has largely eliminated this problem.

Often, uneven bands on slab gels are seen at the anodal end of PAGIF patterns, which are not corrected with larger diameter, or ribbon electrodes. This is normally due to impure or old acrylamide with its various breakdown products. This may be corrected by recrystallizing the acrylamide monomer from chloroform, (see gel preparation Chapter 1). This has been observed (unpublished observations) in aged commercially purified acrylamide that has been used repeatedly for several weeks, even though stored under dessication at -20 °C to -40 °C. This is a problem often blamed on the ampholytes and is more pronounced as one attempts to increase the resolution by increased voltage gradients.

3.7. Additives

3.7.1. Focusing in the presence of detergents

Many of the more popular present methods in PAGIF include some kind of detergent as a necessary part of the system. This area has been the subject of several excellent and extensive reviews, unfortunately too long to be included completely in this section. Hjelmeland *et al* .(47), Hjelmeland and Chrambach (48,49) and Makino (50) have written the most comprehensive of these, which should be reviewed by any one using these methods. It is pertinent in this secton to look at the major rationale for their use in various procedures.

The disruption of hydrophobic interactions, protein-protein or protein-lipid may be required to separate optimally the components in a given system. Detergents are effective in breaking these types of interactions because they provide hydrophobic sites which are capable of binding to hydrophobic regions of either proteins or lipids. Hydrophobic interactions are rather general and may be found in hydrophobic membrane proteins as well as in the soluble hydrophilic proteins. For membrane proteins, detergents are required to disperse the membrane lipids and to render the proteins into a water soluble form. In the water soluble proteins containing hydrophobic domains, detergents provide nonpolar sites with which to interact with hydrophobic regions. For example desoxycholate (DOC) and Triton X-100 may be used as probes for hydrophobic binding sites on proteins. These detergents do not usually denature proteins and expose previouly buried side chains for reaction. They also appear to be weak in breaking protein-protein interactions (49).The important difference between membrane and soluble proteins is the amount of detergent relative to protein which is required to achieve the desired solubilization and or disaggregation.

A primary consideration of choice of a suitable detergent is whether one desires the *native* or *denatured state*; the latter denoting a process by which proteins are unfolded and the tertiary structure modified.

In common usage denaturation also implies several additional phenomena which may have no relationship to changes in tertiary structure. High detergent concentrations may promote the breaking of noncovalent protein-protein or protein-subunit interactions which may be required for the expression of a specific biological activity. An example is in the loss of activity of the insulin receptor (48), where for example, an oligomer of the receptor may be required for effective binding of insulin, and binding of detergent may dissociate the oligomer to the monomer with a resultant loss of binding activity. Thus, while such conditions are not true *denaturation*, *i.e.* an unfolding of the tertiary structure, the original activity is lost. An additional possibility is that the detergent may in some manner act as an inhibitor of the activity without affecting either association or tertiary structure. For example, in the case of cytochrome P-450, detergent at concentrations far below the level required for solubilization is seen to inhibit catalytic activity even though the chromophore is intact (51).

Several attempts have been made to segregate detergents into *denaturing* and *nondenaturing* classes, Tanford (52), Hjelmeland (53), Hjelmeland *et al.* (47), Simonds *et al.*(54), Malpartida and Serrano (55), Vyvoda *et al.* (56) and Coleman (57) *et al.* Unfortunately none of these reports can conclusively tie structure of the detergent to function and although certain head -tail characteristics and groupings seem to play an important role. Further biochemical insight into the interaction of these compounds is required to find a true *nondenaturing* detergent. The more commonly used detergents described by Hjelmeland and Chrambach (48) are given in Fig. 6. Additional characteristics of these detergents utilized in isoelectric focusing are shown in Tables 1a and 1b. (49)

Structural Formula	Chemical or Trade Name
Anionic Detergents	
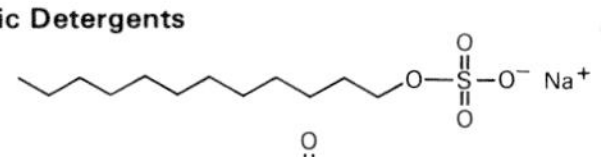 $O{-}SO_2{-}O^-\ Na^+$	Sodium dodecylsulfate (SDS)
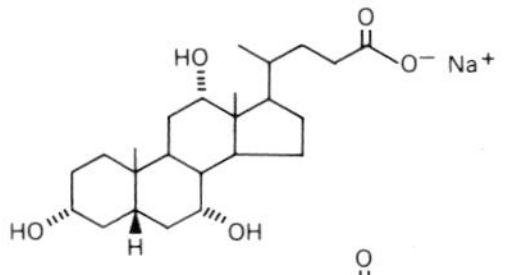 $O^-\ Na^+$	Sodium cholate
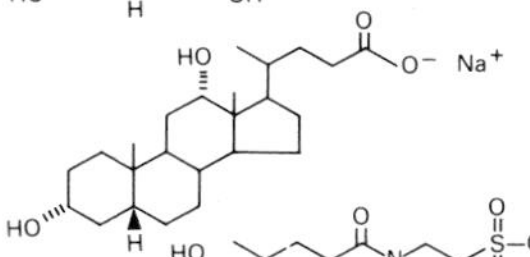 $O^-\ Na^+$	Sodium deoxycholate (DOC)
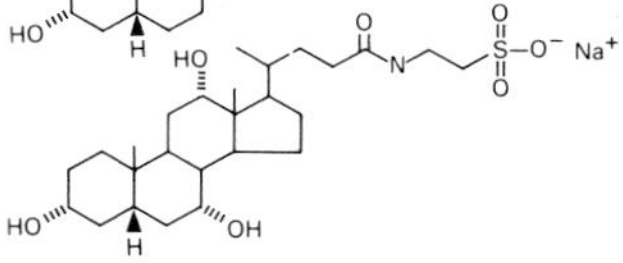 $N{-}SO_2{-}O^-\ Na^+$	Sodium taurocholate
Cationic Detergents	
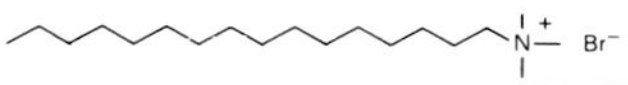 $N^+\ Br^-$	Cetyltrimethylammonium bromide
Zwitterionic Detergents	
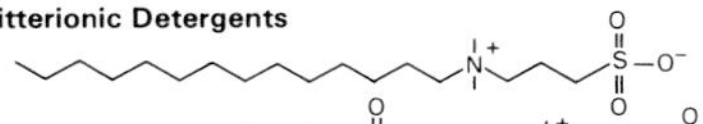 $N^+{-}SO_2{-}O^-$	Zwittergent 3-14
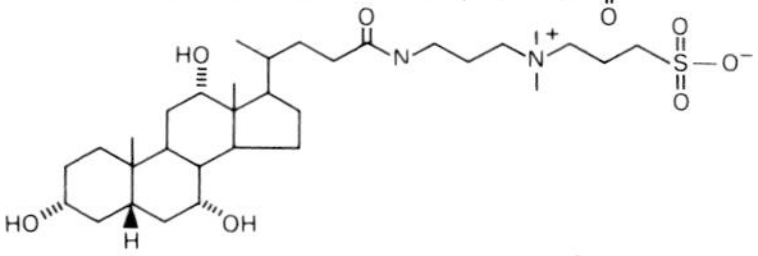 $N{-}N^+{-}SO_2{-}O^-$	CHAPS
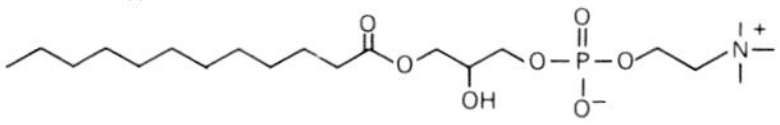 $O{-}P(O)(O^-){-}O{-}N^+$	Lyosophosphatidylcholine
Nonionic Detergents	
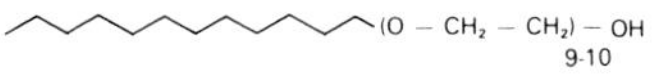 $(O - CH_2 - CH_2)_{9\text{-}10} - OH$	Lubrol PX
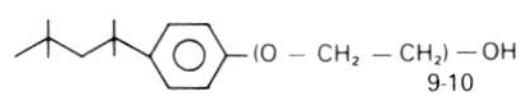 $(O - CH_2 - CH_2)_{9\text{-}10} - OH$	Triton X-100
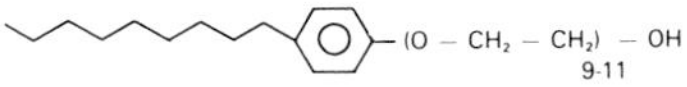 $(O - CH_2 - CH_2)_{9\text{-}11} - OH$	Triton N-101
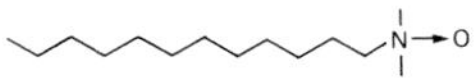 $N \rightarrow O$	Ammonyx LO
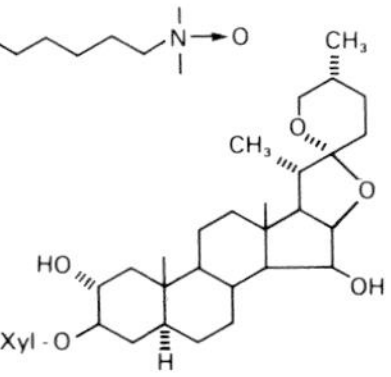 Glu - Glu - Gal - Gal - Xyl - O	Digitonin
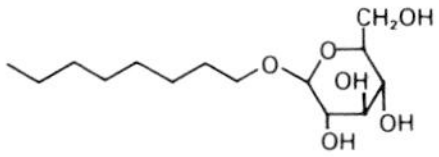 CH_2OH	Octyl glucoside

Fig. 6. Common detergents by class used in electrophoresis (From Hjelmeland and Chrambach) (48)

Table 1a. Characteristics of various common detergents.

	Sodium Cholate	CHAPS	BIGCHAP	Digitonin
Monomer mol. wt.	431	615	862	1229
Micelle Mol. wt.	1,700	6,150	6,900	70,000
Critical micelle conc. % (w/v)	0.36	0.49	0.12	-
Dialyzability	+	+	+	–
Suitable for "Charge Fractionation"	–	+	+	+
Binds divalent cations	+	-	–	-
Significant A_{280}	-	–	– +	-
Interference with protein assays	-	–	+	–

Table 1b. Characteristics of various common detergents (cont.).

	Zwittergen 3-14	Octyl-glucoside	Triton X-100	Lubrol PX
Monomer mol. wt.	364	292	650	582
Micelle Mol. wt.	30,000	68,000	90,000	64,000
Critical micelle conc. % (w/v)	0.011	0.73	0.02	0.006
Dialyzability	-	+	-	-
Suitable for "Charge Fractionation"	+	+	+	+
Binds divalent cations	-	-	-	-
Significant A_{280}	-	-	+	-
Interference with protein assays	-	-	+	-

3.7.1.1. **Urea**

Isoelectric focusing in the presence of urea has been extensively used as a dissociating agent, particularly in the first dimension of two-dimensional electrophoresis after the method of O'Farrell (44). In addition to its dissociating ability, Gianazza *et al.* (58) have suggested that this compound strongly inhibits cathodal drift. There are profound concentration affects of urea on the resoluton of proteins as is demonstrated in the separation of orosomucoid by Altland (59) in Fig. 7, in the second dimension of a 2-1 dimensional separation (see Chapter 4, section 4.2.1.).

Righetti *et al.* (60) have utilized 8 **M** urea in electrophoretic titrations.In this system, many proteins will exist in random coils, subunits will be split, buried groups will be exposed to the solvent and bound ligands or co-factors will be stripped away. Addition of NP-40 to this system helps to stabilize it and to prevent precipitaton.

It should be recognized that separations carried out in urea containing gels can take up to two times as long as those run in its absence. Ultrathin-layer gels, in particular, must be run longer and at a much lower voltage gradient otherwise burning of the gel will result.

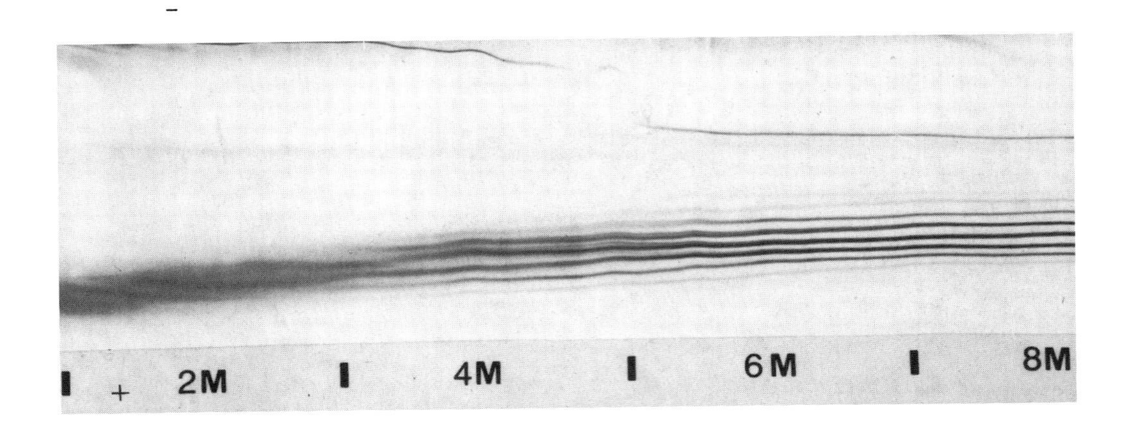

Fig. 7. Isoelectric focusing step in a step gradient of urea perpendicular to the pH gradient of orosomucoid. (From Altland) (59)

3.8 PAGIF Methods in Clinical Medicine

The earlier discontinuous methods, that promised so much with their, for the time, high resolution, have largely been displaced by isoelectric focusing and 2-D electrophoresis. The former normally dealing with native proteins and the latter with both native and denatured materials. Since PAGE is limited in its resolving capacity to some 30-35 proteins, the use of PAGE in this area will not be considered here. For a summary of earlier work in this area, the interested reader is referred to a review by Allen (61).

3.8.1 Standardization of sample collection and handling

Blood sample collection and handling procedures in each hospital are usually processed on a strict routine and this area might not seem to warrant comment. However, the higher the resolution technique that is used the more minor alterations resulting from handling and processing inconsistencies may be magnified, leading to greater variation in results. Many of the tests most applicable to these techniques, thus, must be handled as non-routine, special chemistries. Precautions must be taken assure that the sample arrives at the laboratory as soon as possible after collection.

Serum should be drawn from the blood samples immediately after clotting and the clot is best allowed to contract in the cold. Plasma for enzymes clotting factors, and lipoproteins should be obtained from chilled blood samples and run as soon as possible, or stored frozen at -70 °C, with the exception of samples for lipoprotein analysis, which may be stored at 4 °C for not more than 18 hours. On the other hand, hemoglobin samples may be treated with potassium cyanide and glycerol and held at -20 °C for subsequent analysis for periods of two weeks without affecting separation patterns qualitatively or quantitatively of HbA_{1c} (61). Urine may be concentrated by freeze dialysis and then stored at -70 °C. Long term storage of samples for future reference, or for comparative serial studies, should be divided into aliquots to avoid repeated freezing and thawing and stored at -70 °C or lower.

3.8.2 Standardization of apparatus and operations

The usual clinical requirement for the ability to process a number of samples and controls simultaneously favors the choice of equipment which minimizes the risk of apparatus-inherent complex manipulations. Equipment designed to produce gel slabs offers a number of advantages over those designed for separation in cylindrical gels, or gel rods: 1) On flat slabs many samples, *i.e.*, 12-30 and up to 96 with some systems (63), can be applied side by side and separated under identical conditions; 2) standards may be included on the same gel for sample side-by-side comparison and to act as internal control checks.

3) Flat slabs, as normally used, vary from 0.1 mm to 2.0 mm thick and have more efficient dissipation of the Joule heat produced during separation than do 1.5-3.0 mm diameter cylindrical or rod gels. (The use of capillary tube gels is not considered in this section, since they have had a rather limited application in the routine clinical laboratory).4) Due to their rectangular cross-section, flat gels are better evaluated by quantitative microdensitometry with a much decreased risk of optical artifacts.

3.8.3. Standardization of gels formation and separation conditions

A number of aspects for the control required to achieve reproducible gels prepared "in-house" have been reported (63). Reagent purification and polymerization conditions have been reported by Chrambach and Robard (65). Catalysts and polymer cross-linking concentration effects have been discussed by Watkins and Miller (66) and Righetti (67) (see gel - Chapter 1 section 1.2.).

3.8.4 Plasma proteins

Polyacrylamide gel has experienced considerable interest and use as a clinical tool for the study of plasma proteins. The method has the high-resolution capacity that PAGE seemed to promise, but in the light of experience was found to lack. The lag between research development and routine clinical application with this techinque is perhaps to be expected as with any other newer technique. While apparatus costs are considerably higher than for agar gel and cellulose acetate systems, they are well below those of a number of routine instruments normally found in the clinical chemistry laboratory.

Since the review of Latner in 1975 on PAGIF as a clinical tool (68), numerous advances have been made in the area of clinical pathology and methodology. Banding positions, or isoelectric points, of an ever-increasing number of clinically important proteins have now been identified. Likewise, a number of reports of multiple molecular forms of enzymes from serum, CSF, urine, saliva, and tissues have been published, providing a foundation for the cataloging by isoelectric point of many clinically important body fluids and tissues, enzymes and proteins.

An important advantage of PAGIF on horizontal flat slabs as compared to earlier PAGE is the ability to separate large numbers of samples side by side. Also, the focused sample components are presented at the surface of the gel allowing the application of two techniques which greatly expand the analytical potential of this technique.

The immunoprint technique developed by Arnaud, *et al.* (69), primarily to study the α_1 Pi system, allows separated sample components to be identified immunologically directly from the gel without resorting to a more complex crossed immunoelectrophoresis procedure. The immunoprint technique has been used also to study transferrin and haptoglobin α_1 - Antitrypsin, Gc, IgA and IgG (70), ceruloplasmin and thyroxin binding globulin (71) and vitamin D binding protein (72) heterogeneity. (This procedure is given in detail in the chapter on Methods of Protein Visualization). Any of the many serum proteins for which specific immunological reagents are available, may be directly analyzed with this technique.

Alternatively the reacted PAGIF gel may be washed overnight and the specific immuno-precipitates demonstrates directly in the gel. Precaution should be taken with high molecular weight proteins to assure that they are washed from the gel and that false positive reactions are not presented. This is simply accomplished by using the adjacent track with the same sample to serve as a control to assure that all proteins have been removed and specific proteins remain. Immuno-deletion or subtraction as a first step (73,74) may also be utilized to verify further the specificity of the reaction.

The replicate printing technique developed by Narayanan and Raj (75) and Pretch *et al.* (76), has not only the potential of the application just described, but also the ability to perform a number of additional biochemical tests on each separation. This technique is described in the enzyme visualization section of Chapter 5 (section 5.5). While each print is weaker than the preceding one, up to five or six prints may be made in practice from a given separation. Thus, proteins, glycoproteins, various enzyme tests, and immunoprints can be made from a given gel separation. The exact location of each band is the same on all plates allowing multiple characteristics of each component, or even overlapping components, to be determined simultaneously. The development of commercially available thin gels backed on flexible film by LKB, Serva and Marine Colloid has provided a convenient material to both make printing readily feasible and allow cross-comparison, since the backed master does not shrink or swell during fixation and staining procedures following printing. Backing of gels on other supports such as glass with Silanes is also readily accomplished.

3.8.4.1. Immunoglobulins

PAGIF has been successfully applied in the analysis of the extensive heterogeneity of the immunoglobulins. This technique has been used to study both the heterogeneity resulting from distinct structural genes and that generated by post-synthetic modification of biosynthetically homogeneous protein. With thin-layer PAGIF Williams, *et al.* (77) have shown in pH 5.5–7.5 gradient gels, with a resolving power of 0.005 pH units, that there are some 400 theoretical focusing positions for a single antibody band using the method developed by Awdeh, *et al.* (78). However, impressive as this resolution may seem, Kreth and Williamson (79) have determined that a minimum statistical estimate of 8000 monoclonal antibodies against the haptene 3-nitro-4-hydroxy-5-iodophenylacetate could be produced in the C_3H mouse. In a multiple-band spectrum of monoclonal antibodies, therefore, it is theoretically possible to distinguish 5×10^4 different isoelectric spectra in this pH range (77). As an additional tool in differentiating such spectra Keck, *et al.* (80,81) have developed autoradiographic methods on thin-layer gels where the antibodies are first immobilized by treatment with sodium sulfate and then glutaraldehyde. The bands were then treated by [^{125}I]-labeled antigen and analyzed by autoradiography.

From a more practical clinical viewpoint, PAGIF has been used by Cornell (82) to show that in some instances the method would, like PAGE, detect monoclonal proteins in serum before they were detectable by cellulose acetate or immunoelectrophoresis. Awdeh, *et al.* (83) have shown that the origin of the microheterogeneity from a plasma cell tumor, which produces a single molecular species of IgG_2, is due to lability after synthesis. Brendel, *et al.* (84) have reported also that non-myelomatous monoclonal IgG proteins possess the same individuality and limited microheterogeneity as myelomatous monoclonal IgG proteins.

Trieshmann, *et al.* (85) have analyzed a series of human IgG auto antibodies and have found them to consist of IgA, IgM, and IgG. All demonstrated single peaks by liquid column focusing with isoelectric points from pH 3 to 4.5 in contrast to normal immunoglobulins which focus between 5.5 and 7.5. Assessing sera for such components is also feasible on IgM acid components in large pore gels which would provide a more definitive method to detect microheterogeneity than liquid column isoelectrofocusing.

Dale *et al.* (86) have demonstrated that it is possible to distinguish IgA myelomatosis from IgG myelomatosis using two-dimensional gel techniques and even the IgA light chain can be demonstrated in the serum. Cwynarski, *et al.* (87) have described IgG paraproteins from patients with various gammopathies, both malignant and benign, and present the possibility that a number of "benign-paraproteinemias" may in fact represent premalignant phases. Bouman, *et al.* (88) have investigated the heterogeneity of IgG from plasma cytomas and found the monoclonal IgGs to display 3 to 13 bands. They also indicated that post-synthetic deamidation was not found to occur spontaneously and consequently is not the cause of the heterogeneity observed.

3.8.4.2. β - Globulins

The allotyping of complement components C1, C2, C3, C6, C7 and factors B and D in whole serum separated by PAGIF followed by an in-gel hemolytic assay has been reported by Hobart and Lachmann (89). Polymorphisms in C2 were found to be unlinked to HLA although C2 deficiency is known to be linked genetically to this system.

Transferrin which migrates as a single zone in PAGE has been shown by Hovanessian and Awdeh (90) to be separable into two zones by isoelectric focusing with isoelectric points of 5.6 and 5.2, respectively. The former is monoferric transferrin and the latter diferric transferrin.

Fibrinogen heterogeneity has been demonstrated by Gaffney (91) and Soria, *et al.* (92), who have also employed PAGIF to study abnormal fibrinogens. Arnesen (93) has employed this technique to the study of fibrinogen degradation and fibrin split products.

The diagnosis of Type 3 hyperlipoproteinemia (Broad-β Disease) is based on the demonstration in fasting plasma of a low-density lipoprotein, which in zone electrophoresis migrates in the β-region. Godolphin and Stinson (94) have demonstrated a Sudan Black B-staining band in untreated serum with an isoelectric point of pH 5.44 which appears to be characteristic of the disease. Utermann *et al.* (95) have shown that this disease is further characterized by lack of an autosomal recessive inheritable trait. These authors have also developed a rapid screening method which does not require prior ultracentrifugation of the serum for such analyses.

3.8.4.3. α_2-Globulins

Alpha 2-macroglobulin has been studied by Jones, *et al.* (96) in relation to its subunit structure following thiol reduction. Heterogeneity has been studied by Frenoy and Bourrillon (97), and Ohlson and Skude (98) have demonstrated semi-quantitatively the determination of complexes between various proteases and human α_2-macroglobulins.

Ceruloplasmin may be detected by means of its oxidase activity on p-phenylene diamine, or by using standard immunological reagents for ceruloplasmin.

The haptoglobins may be separated by PAGIF and identified as to genetic type by the addition of hemoglobin with or without enhancement by benzidine staining. Chappuis-Cellier (70) has used the immunoprint technique to localize the haptoglobins. The potential advantage of PAGIF for the study of the haptoglobins is in elucidating greater heterogeneity and subtypes in conjunction with simplified set-up procedures.

The α_2-HS glycoprotein heterogeneity has been observed by Hamberg, *et al.* (99), who reported an isoelectric point range of 4.7-5.1 which is just outside of the pI range of the α_1 proteinase inhibitors but overlaps that of the heterogeneic plasma glycoprotein kininogens. Thus, the α_2-HS glycoproteins may require differentiation from the kininogens by techniques such as the immunoprint method or by selective affinity deletion chromatography or pseudo-ligand affinity chromatography steps prior to isoelectric focusing (43). The group specific (Gc) proteins are discussed in detail in Chapter 4, section 4.3.3.

3.8.4.4. α_1- Globulins

The α_1 Pi system has received considerable interest in the last few years with well over 300 separate reports appearing in the literature. Deficiency states have been associated with a predisposition to emphysema (100), infantile cirrhosis (101), juvenile-arthritis (102), non-allergic asthma (103), and other inflammatory diseases. Allen, *et al.* (104) have found that the more serious heterozygous deficiency states MZ and MS predispose individuals to periodontal disease marked by erosive bone loss as does the M2 allele associated with the A, B, or AB blood groups. Pi typing by PAGIF developed by Allen, *et al.* (105) and Arnaud, *et al.* (106) allows ready differentation of the more than 50 alleles and some 56 phenotypes presently described. The heterogenic system, first demonstrated on the "so-called" *acid starch* gel system of Fagerhol and Laurell (107), is capable of being resolved fully only by PAGIF on a pH 3.5 - 5 gradient or on a pH 4 - 5 Immobiline gradient. Sixty samples can be processed per gel slab within two hours on thin gels and in less than an hour on ultrathin-layer gels. Pi typing is a more accurate and practical way of detecting heterozygote deficient individuals than radial immunodiffusion (RID) and trypsin inhibitory capacity techniques. Since α_1 Pi is an acute phase reactant and serum levels in heterozygote deficiency states are often raised during infection or in patients treated with steroids, particularly estrogens, therefore, individuals with deficiency states could present with serum levels in the low normal or normal range. Over 30 per cent of MS Pi types show low normal or normal levels even in the absence of infection or steroid treatment (108).

Diseases associated with Pi deficiencies are shown in Table 2 on the following page. With the exception of emphysema, infantile cirrhosis and periodontal disease, the link of deficiency states to the diseases listed is somewhat more tenuous and the deficiency itself may not be the primary cause of the disease.

Table 2. Diseases or syndromes in which α_1-antitrypsin deficiency has been reported

Disease or syndrome	Phenotype
Emphysema	ZZ, MZ
Infantile cirrhosis	ZZ
Glomerulonephritis	ZZ
Acne urticaria and angioneurotic edema	Decreased plasma levels, Pi ZZ
Billary atresia	Pi ZZ
Ehlers-Danlos	Pi null
Chronic pancreatitis	Pi MZ and ZZ
Weber-Christian syndrome	Pi ZZ
Spherocytosis	Pi MZ
Thyroidtis	Decreased plasma levels
Gastric or duodenal ulcer	Decreased plasma levels
Severe combined immunodeficiency	Pi ZZ
Coeliac disease	Decreased plasma levels
Psoriasis	Pi MZ
Familial hypercholesterolemia	Pi ZZ
Inherited aminoaciduria	Pi MZ and ZZ
Multisystem fibrosis	Pi SZ
Fibrosing alveolitis	Pi MZ
Multiple endocrine adenomatosis	Pi ZZ
Pyloric stenosis and hyperbilirubinemia	Pi ZZ
Allergic contact dermatitis	Pi MS and MZ
Uveitis	Pi MZ
Periodontitis	Pi M_2 with A,B or AB blood group and Z, S, and P

3.8.4.5. α_1 Lipoproteins

Separation of lipoproteins on gel slabs in which whole serum is placed on filter paper tabs results in considerable surface smearing of the VLDL and LDL which do not penetrate the gel. Gel rods containing a five percent monomer concentration have been used by Kostner, *et al.* (109) to separate the lipoproteins with the HDL region showing four distinct bands. Kostner, *et al.* (110), utilizing a modified system employing photopolymerized gel rods containing 23 percent ethylene glycol, were able to separate eight bands from prestained sera. While HDL_2 and HDL_3 heterogeneity has been demonstrated by sucrose gradient column studies by Sodhi, *et al.* (111), Blaton, *et al.* (112) and Eggena, *et al.* (113), PAGIF has not been widely used. Removal of VLDL and LDL from serum by heparin and magnesium as described by Burnstein and Scholnick (114) will provide samples on which HDL may be separated without resorting to prior density gradient ultra-centrifugation. Scanu, *et al.* (115) have studied the HDL in depth by analytical PAGIF and have found that focusing is most valuable in the characterization of products separated and purified by prior chromatographic procedures and Zannis and Breslow (116) have determined the isoelectric point of the apo-lipoproteins.

The thyroxin-binding globulin has been shown by PAGIF to exhibit heterogeneity similar to the other α_1-globulins. Some 9 bands have been found with PAGIF by Marshall, *et al.* (117), with 4 major and 5 minor bands from purified material obtained from pooled plasma with a pI region from 1.2 to 5.2. Although this reported range of isoelectric points overlaps that of the proteinase inhibitors, the low concentration of thyroxin binding globulin, 1-2 mg per 100 ml serum, in practice, causes no interference with Pi typing. Detection of the thyroxin-binding globulins by immunoprint techniques may be difficult due to their very low concentration but may be enhanced by the more sensitive silver stains and separated for α_1 by selective affinity deletion techniques prior to PAGIF (43).

Alpha - fetoprotein, normally undetectable within days after birth, reappears in the serum of patients with primary liver carcinoma and embryonal carcinoma. It has been isolated and its heterogeneity studied by PAGIF. Two major α-fetoprotein peaks with isoelectric points of 5.08 and 5.42 have been described by Alpert, *et al.* (118) in both hepatomas and fetal serum. Two compounds of α-fetoproteins have also been described in the ascites fluid from a patient with primary carcinoma by Sokolov, *et al.* (119) with isoelectric points of 4.78 and 5.20, respectively.

3.8.4.6. Hemoglobin

Aside from their clinical importance, the hemoglobins hold a special place in the development of isoelectric focusing. They have played perhaps the most important role in the early development of the technique. As proteins bearing a chromophoric heme group, they were used extensively to demonstrate not only the resolving power of isoelectric focusing, but also as markers to indicate completion of pH gradient formation. They have also served as a model system for subunit exchange, binding studies and cooperativity as reported by Park (120). High-voltage, controlled low-temperature systems have provided a considerable advance in the study of the hemoglobinopathies as well as in the study of human hemoglobin phenotypes as demonstrated by Bunn (121), while Drysdale, *et al.* (122) have compared human and animal hemoglobins.

Chromatographic techniques have been previously used to study heterogeneity of both normal and abnormal types (123,124), but the ability to separate rapidly multiple samples by PAGIF on pH 6-8 gradient gels for both quantitative and qualitative studies offers a valuable tool for clinical studies. Jeppsson and Bergland (125) showed that thin-layer PAGIF was able to separate Hb-Malmo from Hb-A, and Hb-F from $Fb-F_{11}$, which was not possible with previous zone electrophoretic techniques. Another high oxygen affinity hemoglobin with a mutation at B97 and associated with familial erythrocytosis was reported using this technique by Taketa, *et al.* (126). Monte, *et al.* (127) have recently mapped hemoblobin mutants by PAGIF. Altland (63) has developed a mass screening process for hemoglobinopathies using a portion of the dried blood samples obtained for PKU screening. The samples dissolved in 50 pl of 0.02**M** KCN were applied in two rows of 48 samples to a single gel. A center anode strip is used with a cathode strip at both the top and bottom of the gel so that 96 samples per run may be analyzed.

Koenig, *et al.* (128) have shown that the measurement of Hb-A_{1c} provides a reliable index of control of carbohydrate metabolism in diabetes, and Trivelli, *et al.* (129) have reported that Hb-A_{1c} is increased two-fold or more in patients with poorly controlled diabetes mellitus. Spicer, *et al.* (62) using high-voltage, constant-power PAGIF have shown that Hb-A_{1c} and the other fast hemoglobins Hb-A_{1b} may be separated and accurately quantified by microdensitommetry on pH 6-8 gradient gels as illustrated in Figs. 8a and b. This method is rapid with the added advantage that abnormal hemoglobin phenotypes are also detected if present. For pediatric patients, these authors have utilized standard heparinized microhematocrit tubes to obtain blood from a finger or ear prick for analysis.

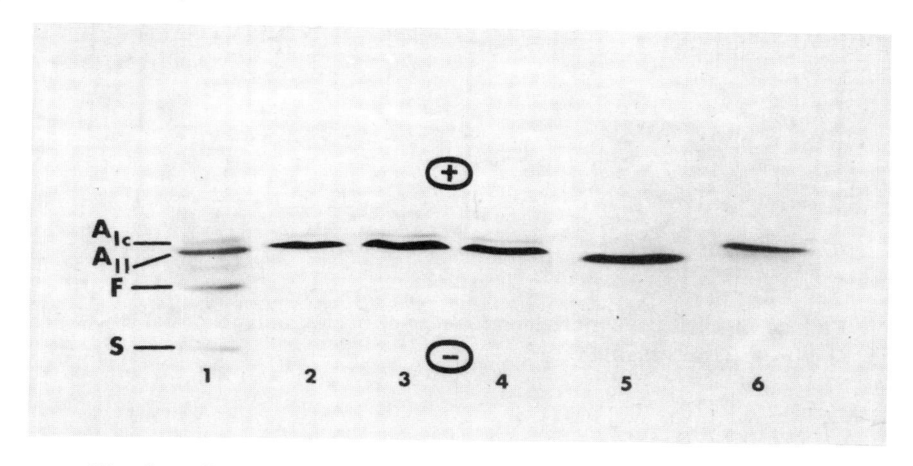

Fig. 8a. Separation of hemoglobin by PAGIF on pH 6.0-8.0 pradient. Sample 1 is a mixture of F,S, A_{II}, and A_{Ic}. Sample 2 is form a normal individual; Samples 3 and 4 from diabetics. Sample 5 is twice rechromatographed, AII from a BioRex column and Sample 6 is similarly rechromatographed HbA_{Ic}. Separations were directly fixed in 12.5 per cent TCA and subjected to quantitative microdensitometry as shown in Fig. 8b. (62)

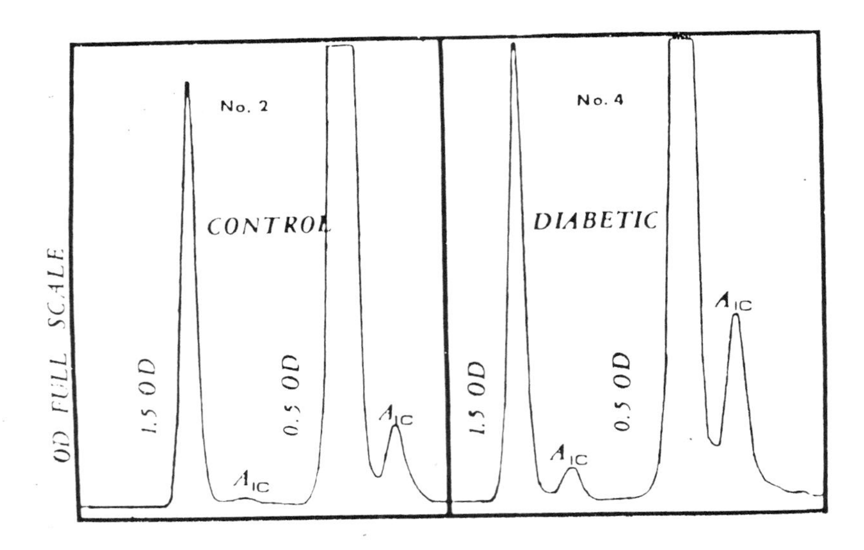

Fig. 8b. Quantitative microdensitometry of normal and abnormal levels of HbA_{1c}. (from Allen) (61)

3.8.4.7. Cerebrospinal fluid

Vesterberg (130) has shown that patients with multiple selerosis have a relative increase in the CSF of alkaline globulins, *i.e.* those with an isoelectric point greater than 7.5 at 10 °C. Delmotte (131) has also reported similar findings. Stibler and Kjellin (132) have demonstrated the focusing patterns in CSF of degenerative neurological diseases. The same authors have also studied the protein patterns in hereditary ataxia, hereditary spastic paraplegia, muscular dystrophy, and spinal muscular atrophies. The low protein concentration of CSF has required concentration with its inherant denaturing problems. However, ultrathin-layer isoelectric focusing in conjunction with the silver stain has been utilized with unconcentrated CSF with good results (133,134).

3.8.4.8. Urine

Robtol (135) appears to have first applied isoelectric focusing to human urinary proteins on polyacrylamide gel. He found some 30 protein zones to be present in normal urine. Vesterberg (130) has identified the location of the β_2-microglobulin and found more than 40 protein zones in urine from patients with advanced renal damage. Hall and Vesiljevic (136) have characterized further the β_2-microglobulin and described two homologies, one with a pI of 5.3 and the other with a pI of 5.7. This portion, which has a weight of 11,500, has been established to be the smaller of the two polypeptides of the HLA antigen and is elevated in the majority of myeloma patients. The highest β_2-microglobulin levels are found in patients with tubular deficiencies and in patients following renal transplantation. Patients afflicted with Balkan nephropathy are a special example of renal tubular disorder in which high levels of β_2-microglobulin are found (136).

Boulton and Huntsman (137) have used PAGIF to distinguish myoglobin from hemoglobin in non-fresh urine of kidney donors where conversion of myoglobin to met-hemoglobin makes spectroscopic regognition of this pigment unreliable. Hultberg *et al.* (138) have studied the enzymes of four acid hydrolases in both kidney and urine and have found that there was a predominance of isozymes with a low isoelectric point in the urine. In contrast, in kidney tissue extracts isozymes with higher isoelectric points predominated. Keller *et al.* (139) have studied the colony stimulating factor from human leukemia urine and found a glycoprotein demonstrating marked heterogeneity with isoelectric points between pH 3.3 and 4.1.

3.8.4.9. Salivary proteins

Beeley (140) has employed PAGIF in gel rods to separate the salivary proteins and Benninek and Cornell (141) and Allen *et al.* (74) have used the method as a criterion in the purification and partial characterization of four proteins from human parotid saliva and raw saliva respectively. Chisholm *et al.* (142) have found additional protein bands with acid isoelectric points in the saliva of individuals with Sjogren's syndrome and those with rheumatoid arthritis. Pronk (143) found three different protein type patterns in parotid saliva which were shown by family-studies to be phenotypic expressions of one autosomal locus with two co-dominant alleles. These were found to be correlated with α-amylase activity. Bustos and Fung (144) have found two proteins in the saliva of cystic fibrotics not found in normal persons.

3.8.4.10. Enzymes

The significance of enzymes present in the plasma, originating from healthy and diseased tissue as indicators of disease and therapy control is readily apparent in clinical medicine. Implicit in such correlations of enzyme levels with disease is that their elevation or decrease is representative of events occurring in a particular disease state at the cellular level. Alterations in the level of a particular enzyme or isoenzyme may result either from the tissue or from a deficit in clearance. Moreover, genetic and physiological factors can singly, or in combination, alter plasma levels in the absence of disease.

Many enzymes may be analyzed by PAGIF using suitable histochemical stains, a number of which have been compiled by Righetti and Drysdale (1), and Maurer (145). PAGIF requires the gel to be pretreated in a sufficiently concentrated buffer to equilibrate the gel to the proper pH for a given reaction. However, gels, particularly ultrathin-layer gels, with ampholyte concentrations of 20-50 **mM** make it quite easy to utilize a 100-200 **mM** buffer directly with the substrate and complexing agents. Wadstrom and Smith (146) have reported ways to overcome adverse pH conditions in PAGIF gels. Additionally, enzymes may loose activity after PAGIF due to chelation of necessary metal cofactors. For exmaple, Latner, *et al.* (147) have reported 90 per cent loss of activity of alkaline phosphatase unless the zinc cofactor was added back to the reaction mixture.

3.8.4.11. Tissue proteins and enzymes

PAGIF has not been widely used as clinical tools to study soluble proteins and enzymes from tissues or cells. However, this potential deserves mention as it is certainly within the microanalytical capability of the system. For example, only a portion of wedge biopsy material is necessary to provide sufficient material for soluble protein extraction. Studies such as the detection of carcino-fetal liver ferritins by Alpert, *et al.* (148). The multiple molecular forms of tyrosinase in melanomas have been described by Burnett and Seilor (149) and tartrate resistant acid phosphatases from leukocytes in hairy-cell leukemia by Yam, *et al.* (152).

Latner (68) has also shown differences in kidney tissue in renal carcinoma and in the cervical mucosa of pap-smear-positive women. Hennis, *et al.* (151) have shown the effect of analgesics on rabbit kidney esterases and Allen, *et al.* (152,153) have demonstrated the effects of cisplatin compounds on the kidney esterases of mice and the reversal effects of diethyldithiocarbamate on platinum toxicity. The studies of Saravis *et al.* (154) and Thompson *et al.* (155) on the direct focusing of frozen tissue sections on agarose is an important and novel technique for both enzymes and proteins of biological interest. A number of these are discussed at length in chapter 5 - Sample Visualization.

Some selected clinically significant enzymes which may be qualitatively and quantitatively analyzed with isoelectric focusing are given in Table 3 along with their diagnostic significance and source of sample material.

Table 3. PAGIF clinical zymogram methods

Enzyme	Tissue	Application	Refs.
Acid Phosphatase	WBC Extract	Hairy cell leukemia	(149, 156)
Alkaline Phosphatase	Serum	Lymphatic leukemia, infectious mononucleosis	(157)
Amylase	Saliva, urine and serum	Rise in pancreatic disease and injury	(158)
Acid α-glucosidase	Muscle, liver	Pompe's disease	(159)
α-L-fucosidase	Liver	Mucopolysaccharoidosis F	(160,161)
Gluc-6-PO_4 dehyd.	Red cells	Genetic variants, hemolytic episodes	(162,163)
N-Acetyl-β-D-hexose-iminidase	Serum, tissue and amniotic fluid	Tay-Sachs	(164-166)
β-Galactosidase	Liver	Mucopolysaccharoidosis Type 1, 2, 3	(167)
Esterases	Skin, serum, kidney, spleen	Wound esterase, bacterial and viral pneumonias, Platinate drugs, acetometaphin and histiocytic neoplasms	(151,152) (168-170)
Tyrosinase	Tissue	Melanoma	(148)
Peroxidase	Red Cells, uterine fluid	Estrogen induced growth	(171)
Creatinine phosphokinase	Serum	Myocardial injury	(172,173)
Lactic dehydrog.	Serum red cells,urine	Myocardial injury, pyelonephritis	(174-175)
Dipeptidyl amino peptidase	Macropages	Murine Pnemonia	(176)
Pyruvate kinase	red cells	Hereditary hemolytic anemia	(177)

3.9. Genetic and Forensic Applications

The resolving capability of isoelectric focusing makes it a valuable method to study genetic polymorphisms and it employment in this area of investigation has followed almost a geometric progression of use. This field and its associated use in forensic applications has grown so greatly as to warrant a book on each subject alone. In this section, therfore, we will only attempt to give an overview of some of the more common applications, using the forensic field as examples presented in tabular form and several representative illustrations.

Gentically controlled polymorphisms expressed at the molecular level detected by isoelectric focusing are, perhaps, best exemplified by the hemoglobins. Aside from their clinical importance, they evidence over 300 genetic variants, most of which are not associated with any clinical disease. They, therefore, serve admirably for both genetic and forensic purposes. An added advantage is that they may be differentiated directly from the same filter paper spots used for PKU testing and thus lend themselves to mass screening as described by Altland (63) and illustrated in Fig. 9.

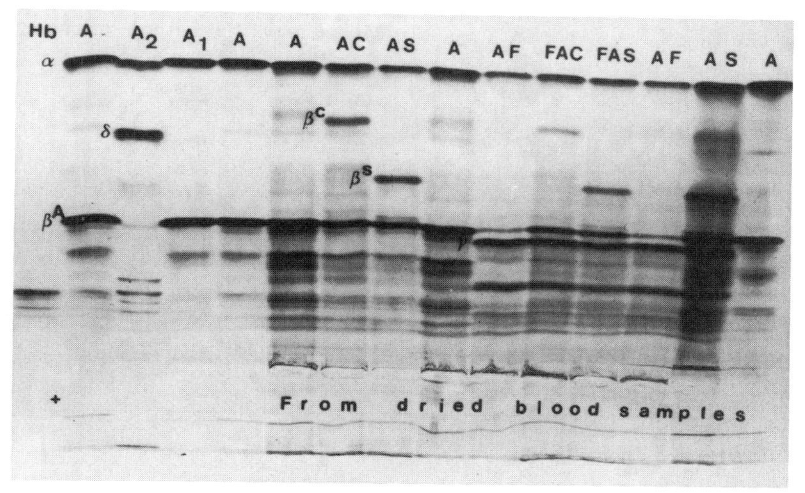

Fig. 9. Isoelectric focusing of human globin chains (α , β A-, βS-, βC-, γ and δ- chains from dried blood. (from Vogel and Altland) (178).

Table 4. Enzymes commonly used in forensic medicine.

Enzyme	Source	Method	Substrate	Ref
Acid erythrocyte phosphatase	RBC, semen, prostate	Direct	4-methyl - umbellferyl phosphate	(179)
Adenosine deaminase	RBC	Indirect	FAD, MTT	(179).
Alkaline placental phosphatase	Placenta	Direct	β-Naphthyl phosphate	(180)
Alcohol dehydrogenase	Serum	Indirect	NADP, PMS	(181)
Alpha-1 fucosidase	Leukocytes	Direct	4-methyl-umbelliferyl-α-L- fucoside	(182)
Carbonic anhydrase	RBC	Indirect	NADP, PMS, MTT	(180)
Erythrocyte acid phosphatase	RBC	Direct	β-Naphthyl phosphate	(180)
Erythrocyte NADH-diaphorase	RBC, leukocytes, thrombocytes,	Direct	DCIP, MTT	(183)
Esterase D	RBC	Direct	4-methyl- umbelli-feryl acetate	(184)
Glyoxalase I	RBC	Indirect	DCIP, MTT	(185)
Glutamic pyruvate transaminase	RBC	Indirect	$NADH_2$, MTT	(186)
Phosphoglucomutase I	RBC, Leukocytes,	Indirect	NADP, MTT	(187)
6-Phosphogluconate-dehydrogenase	plasma	Indirect	NADP, MTT	(188)
Superoxide-dismutase A (Indophenoloxydase)	RBC	Indirect	O-dianisidine	(190)

Table 5. Protein polymorphisms commonly used in forensic medicine.

Protein	Source	Detection	Ref
Alpha-1-Antitrypsin	Serum, Saliva Sweat	Coomassie Blue, immunoprint silver, Immobiline	(105,191,192)
Antithrombin III	Plasma	Coomassie Blue, immunoprint	(193)
Gc proteins	Serum	Commassie Blue, Immunoprint silver	(194)
Haptoglobin	Serum	Hemoglobin binding Diaminobenzidine	(195)
Hemoglobin	RBC	Direct, diaminobenzidine	(196)
Transferrin	Serum, sweat	Coomassie Blue, Immunoprint silver	(90,191)
Orosomucoid Acid $\alpha_1$glyco- protein	Serum	Coomassie Blue, Immunoprint silver	(197,198)
Apolipoprotein E	Plasma	Coomassie Blue, Immunoprint	(199)
Complement factor C2, C4, C6 and C7	Serum	Coomassie Blue, Immunoprint	(89,200)
Properdin factor B	RBC	Coomassie Blue	(201)
Factor D	Serum	Coomassie Blue	(89)

The application of genetically controlled polymorphisms is also of great importance in the agricultural community, where its ability to recognize and distiguish between different cultivars (varieties) of particular crops is essential in the selection of optimal varieties for a number of purposes, ranging from the making of bread to the brewing of beer and wine. In those countries where seed merchants are required by law to supply seed of the correct cultivar, this is the only way to assure seed purity prior to actually growing the plant. Cooke (202) has recently extensively reviewed this field, covering not only the application of isoelectric focusing, but also all electrophoretic methods commonly accepted in this area. Righetti and Bianchi Bosisio (203) have reported the applications of isoelectric focusing in the analysis of plant and food proteins from pasta to potatoes. In fact, in the latter Stegemann and Loeschke (204) have compiled an *Atlas der Kartoffelsorten auf der Basis von Gel-Elektrophoresen*. Drawert and Görg (205) have also described the protein profiles of a number of different varieties of grapes, important in the production of another important German product (white wine) using isoelectric focusing.

The detection of adulteration in foods is another area in which isoelectric focusing can play an important role and as an example of this potential we have chosen an area important to the region of two of the authors, the detection of species of fish (what better to go with the white wine above?) illustrated in Fig. 10 .

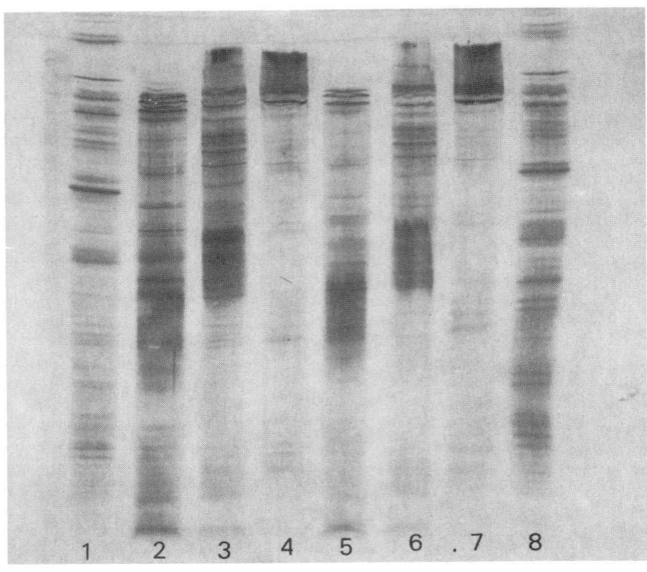

Fig. 10. Isoelectric focusing of flounder extracts on a pH 3 to 7 gradient on a 250μ thick gel for 626 V/h with a maximum of 430 V/cm and stained with diammine silver. Tracks 1 and 8 1.0μl of a 0.1 % Rohament P control, 2-4 1.0μl W, S and WP respectively and tracks 5-7 number 2-4 at a 0.5μl application. (from Allen and Reeder) (unpublished data)

3.10 Biochemical Applications

3.10.1 Titration curves

Rosengren *et al.*(206) reported on a simple method for selecting optimum pH conditions for electrophoresis. This in fact was a direct titration of all the proteins in a sample, which detected the isoelectric point of each. The method consisted of establishing a pH gradient on the gel slab without the application of any sample. The electrode wicks and underlying gel were then removed and a slot cut vertical to the pH gradient at the center of the gel and new electrode wicks containing the same anolyte and catholyte were added perpendicular to the pH gradient at both sides of the slab. Sample was then added to the slot and electrophoresis carried out at 30 V/cm. (While this system may be considered by some to be a two-dimensional technique, the authors have chosen to include it in the isoelectric focusing chapter, since it is closely related to the following section). The setup for this system is presented diagramatically in Fig. 11.

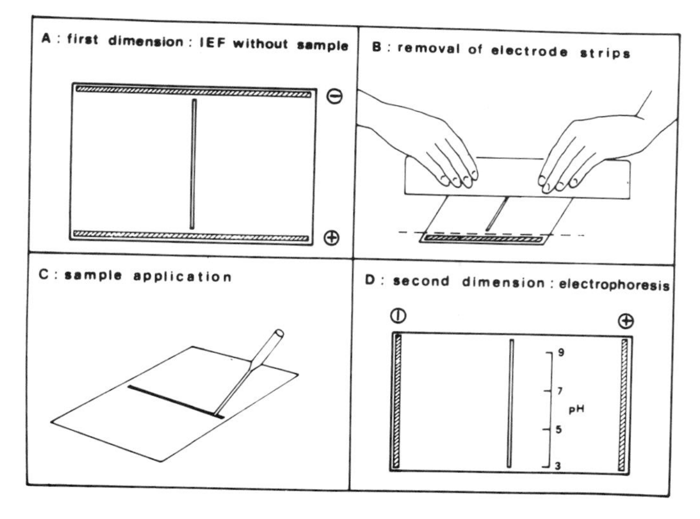

Fig. 11. Experimental setup for generating titration curves by PAGIF in the first dimension and then electrophoresis perpendicular to the pH gradient. (from Righetti and Gianazza) (207)

This technique also allows the differential titration of a protein and its genetic mutants. Righetti *et al.* (209) and Righetti and Gianazza (210). They also showed that the binding of the ligand produced conformational changes in the apoprotein. Molecular interactions of macromolecules have been reported between cytochrome B_5 and methemoglobin by Righetti *et al.* (211). Constans *et al.* (212) have used this technique to demonstrate the binding of vitamin D_3, and its derivatives, to the human serum vitamin D binding-protein. The titration curve of C reactive protein, shown in Fig. 12, has been determined by Laurent *et al.* (213)

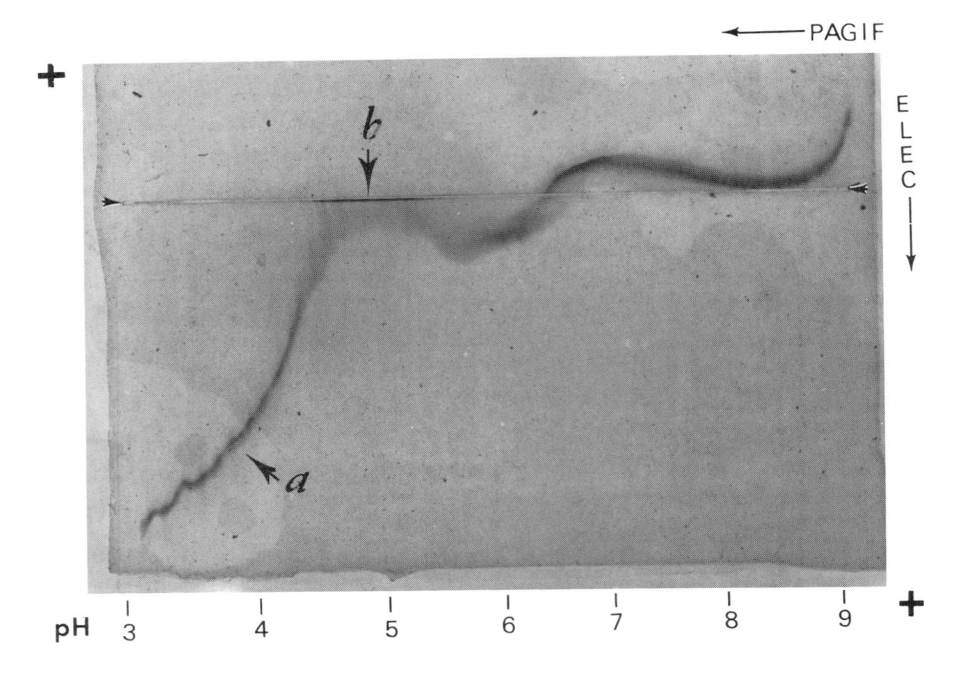

Fig. 12. Titration curve of human purified CRP. The sample load was 220 μg protein in 100 μl sample volume. The arrows and plus signs represent the direction of PAGIF and electrophoresis. The arrowhead indicates the sample application trench and the arrowa at a and b respectively shows a region of dissociation and denaturation. (from Laurent *et al.*) (213)

Righetti *et al.* (207) and Valenti *et al.* (214) have developed a mathematical relationship that allows direct p*K* determination of ionizable groups from the shape of the pH mobility curves. A more recent extension of pH mobility curves is that of determining dissociation constants (K_d) of ligands to proteins and their pH dependence. This stems from the developments of affinity electrophoresis by Horejsi (215). When the ligand to be tested is a large macromolecule it is simply entrapped in the gel, and when a small molecule, it is covalently bound to the gel fibers. In the presence of increasing concentrations of ligand, the titration curve of the protein is progressively retarded, dependent on pH. When the mobility increments are plotted against the molarity of the ligand in the gel, the K_d value can be calculated. Titration curves have a great value for investigating a number of physico-chemical properties of proteins and their interactions with specific ligands. A further extension of this technique in a single dimension of isoelectric focusing is given in the next section.

3.10.2. Affinity titrations

At the same Hamburg meeting where titration curves were introduced by Rosengren *et al.* (206), Allen *et al.* (42) described a technique for the determination of the optimal pH of trypsin binding to alpha$_1$-antitrypsin which employed a single dimension technique, where the gradient was first established and the the sample then added on a diagonal to a prefocused gel for the study of pH effects on trypsin binding to alpha$_1$-antitrypsin. In this case, since the pI of trypsin is above pH 10, it was added 1 cm anodally to the diagonally applied sample, containing the alpha$_1$ -antitrypsins, with pIs of the allele products pH 4.4 to 4.7, and the two allowed to cross focus with the various predetermined pH values. With strongly binding ligand interactions such as trypsin - antitrypsin, the short reaction time during the crossing phase, does not allow the enzyme to react with other proteins for extended periods, with potentially resulting hydrolysis. The gel temperature held at 4 ^{0}C along with the short contact time during crossed-focusing diminishes this problem. Such a system is shown schematically in Fig. 13. and the titration curve of affinity binding of trypsin-antitrypsin is shown in Fig. 14.

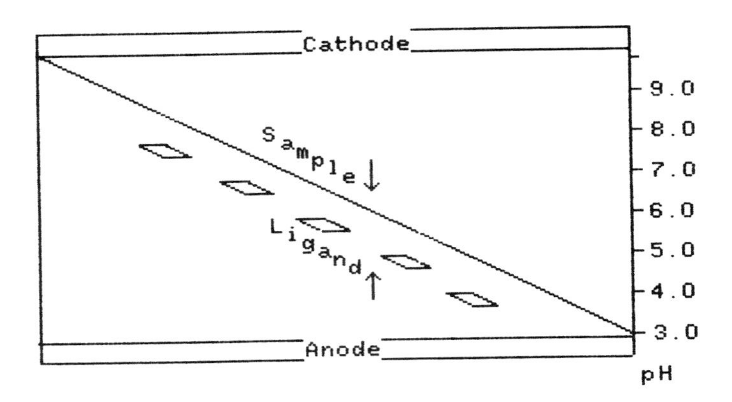

Fig. 13. Diagramatic setup of crossed affinty titration of trypsin and α_1-antitrypsin. The arrows mark the direction of migration.

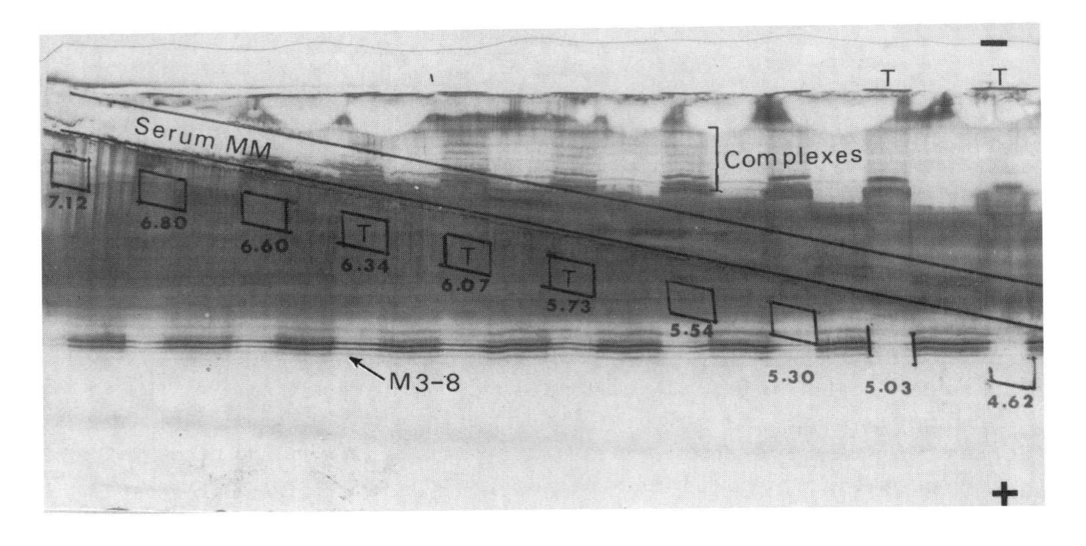

Fig. 14. The pH effect on typsin binding by alpha$_1$-antitrypsin between pH 4.6 and 7.1. Unbound trypsin T is focused at the cathode wick and the complexes formed are marked by the bracket. Alpha$_1$-antitrypsin allele products are marked by arrows. (from Allen *et al.*) (42).

Maximum binding, marked by the diminution of the protease inhibitor, is evident at both pH 4.6 and 7.1. Minimal binding is seen from pH 5.7 to 6.3. By varying the molar concentration of the trypsin and measuring the dimunition of the proteinase inhibitor bands by quantitative microdensitometry, binding capacity and stoichiometric studies can also be carried out as shown in Fig. 15.

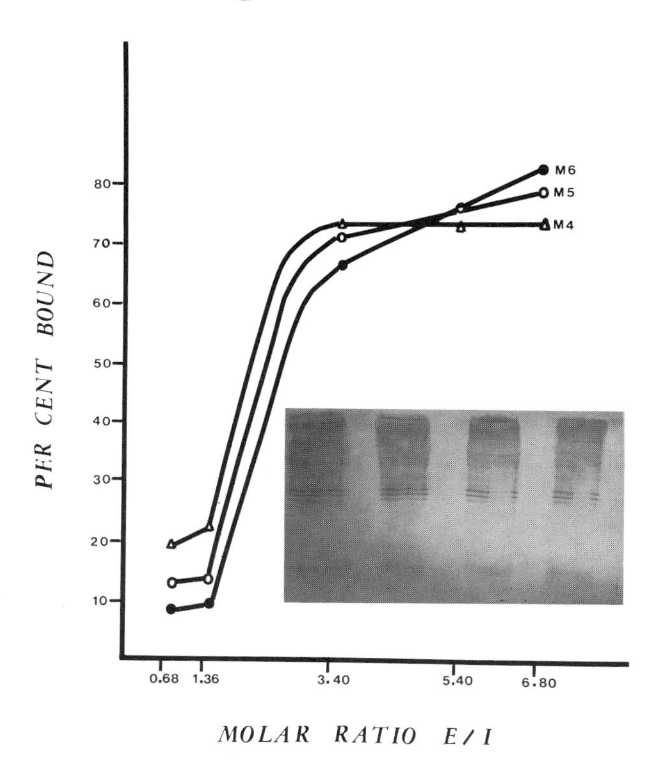

Fig. 15. Binding capacity of varying concentrations of trypsin at pH 7.0 of three M allele products.(form Allen *et al.*) (42)

This technique may also be used to determine proteinase inhibitors in a complex mixtue of proteins by analysis of deleted bands between a control and a crossed trypsin sample as illustrated in Fig. 16.

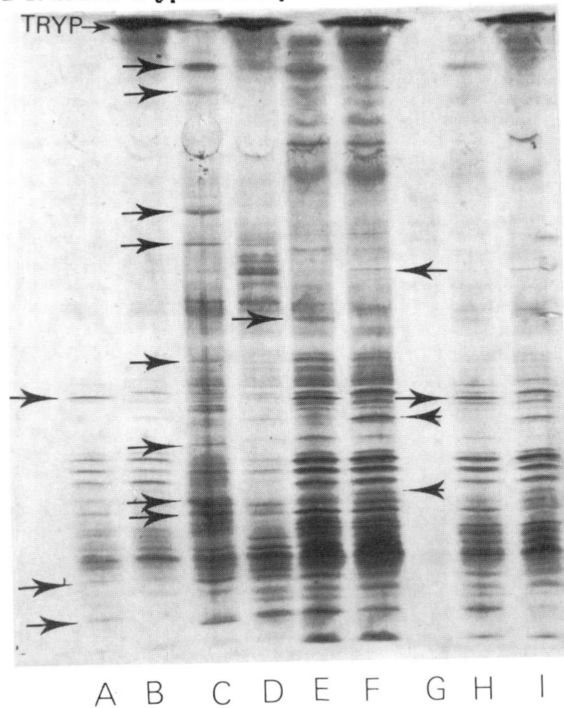

Fig. 16. Protease binding proteins present in lymphocyte extracts (A,B) and granulocyte extracts (C–I). Arrows pointing to right indicate trypsin-binding proteins and arrows pointing left indicate complexes. (from Allen and Arnaud) (216)

Ligands which do not cause proteolysis may be mixed directly with the sample, or if the pI of the protein and ligand are known the ligand can be placed anodal, or cathodal, to the protein and one allowed to overrun the other to effect binding. These procedures in conjunction with those in the last section should allow a wide variety of testing of the physico-chemical characteristics of a given protein, or as an aid in the location of an unknown component.

3.11. Preparative Isoelectric Focusing

Among charge fractionation procedures, preparative isoelectric focusing is particularly advantageous due to high resolution and load capacity (4, 217-221). Depending on the fractionation scale used, preparative isoelectric focusing can be classified into two categories: 1) Techniques for laboratory-scale fractionation of milligram quantities with one gram as an upper limit. 2) Techniques for large-scale fractionation of gram or more amounts. The latter category consists of numerous and diverse systems, e.g. discontinuous (219,220,222). Continuous (218) and recycling (223) systems. which to date, have for one reason or another, not been applied to many practical problems. Most applications of preparative IEF have been described for laboratory-scale separation. The two most commonly used techniques comprise density gradient columns and layers of granulated gels.

Righetti (2) has reviewed preparative isoelectric focusing in sucrose density columns and in continuously flowing density gradients. This section will be limited to flat slab systems using granulated gels simply because, in the authors experience, these methods provide optimal resolution with simplified equipment and ease of set up and sample recovery. We have exempted column systems, since cooling efficiency is poorer limiting power imput and thermal gradients across the column are produced, which tend to limit resolution capability. We have chosen to present a preparative IEF system on granulated gel, since it has the resoluton of most analytical polyacrylamide or agar systems. The system described by Frey and Radola (224) achieves analytical resolution of up to 1g amounts of protein with a resolution of 0.01-0.015 pI. This resolution is achieved with 1) gel layers of reduced thickness (1mm), 2) high field strength in the final stage of focusing (100-200 V/cm), 3) long separation distances (20-40 cm), 4) high volt-hour (V/h) products ensuring steady state conditions. Optimum load capacities for such a system are 3-5 mg protein/ml gel-bed volume and 10 mg protein/ ml gel-bed volume are well tolerated. The operational procedures of this technique are given in Table 6.

Table 6. Preparative focusing on granulated gels

1. Preparation of gel suspension	Sephadex G-200 superfine, or enzymatically resistant Bio-Gel P-60 for use with cellulases and hemicellulases, are suspended in distilled water and the swollen gel washed on a G 4 sintered glass filter with 20 volumes of distilled water. To avoid mechanical damage of the gel beads, the water is overlayered without stirring the gel. After washing, the resulting suspension may be kept at 2 - 4 °C for up to several months without the addition of preservatives.
2. Coating plates	The required amount of gel suspension, with an excess of water,approximately 20 %, to compensate for water loss on drying, is calculated for a 1 mm gel layer from the size of the plate to be used. Carrier ampholytes of the desired pH range are added to a final concentration of 2 % w/v and after shaking for 5 min the suspension is deaerated for a few minutes under *vacuo*. Glass plates 1 mm thick, cleaned by washing with detergent, rinsed and then rinsed in 95 % ethanol and wiped dry are prepared. The gel suspension is spread over the plate with the aid of a glass rod and 1 mm thick guide rails. A perfectly even gel layer with an excess of liquid collecting on the surface is obtained by tapping the plate. The surface is dried with a fan until irregular 1-3mm cracks appear at the edges of the gel.
3. Anolyte and catholyte	One cm wide electrode strips are soaked with the following solutions: Anolyte (0.025 M aspartic acid and 0.025 M glutamic acid); catholyte (2 M ethylene diamine containing 0.025 M arginine and 0.025 M lysine). Take care to establish good contact of the strips to the gel layer and to prevent an excess of buffer from the wicks. Adjustable platinum ribbon electrodes (Desaga or Bio-Rad) are placed on the electrode wicks and weighted with a 4mm thick glass plate.
4. Sample application	The sample is applied as a streak on the gel surface with a microscope slide edge or the edge of a rectangular plate 2 cm narrower than the gel plate to prevent edge effects. Approximately 30 µl of sample solution can be applied per cm width of applicator. Commercial sample applicators (Bio-Rad, LKB Pharmacia and Desaga) allow greater sample loads to be applied in 0.5 -1cm intervals. The sample should be applied one-fourth of the total separation distance away from either electrode, preferably in an area where no components of interest will focus. Two to 10 % protein solution, prepared by ultrafiltration may be applied.

Table 6. Cont.

5. **Focusing**	Typical running conditions for different separation distances are as follows: 1] 10 cm - prefocusing at 50 V/cm for 20 to 30 min followed by a final field strength of 150 V/cm to a total of 1500 Vh; 2] 20 cm gel - prefocusing at 20 to 50 V/cm with a final field strength of 150 V/cm to a total of 10,000 Vh; 3] 40 cm gel - prefocusing at 20 to 40 V/cm for 16h with a final field strength of 100 to 125 V/cm for a total of 30,000 Vh. With adequate cooling 0.05 - 0.1 Watt per cm^2 are tolerated well. Also increased power levels may be employed with higher efficiency heat transfer plates such as beryllium oxide ceramic plates in conjunction with Peltier cooling derscribed in a following section
6. **Band location printing**	Sartorius cellulose acetate membranes 1.5-2 cm wide of appropriate length for the gel are rolled from one end onto the gel layer containing the focused proteins. The membrane is gently pressed onto the surface to assure complete wetting, with care being taken not to entrap any air bubbles. After contact for one min, the membrane is removed and placed for 1-2 min in 20 % TCA. Any adhering gel granules are washed off during this step. The strips are then stained in 1 % Ponceau S and 10 % TCA for 1 min and destained for a few seconds in 5 % acetic acid. The prints are dried in air yielding a pattern on a clear white background.
7. **Removal of ampholytes**	Carrier ampholytes are removed from the gel electrophoretically through a dialysis membrane. (see Fig. 17.) A sheet of filter paper (Whatman # 3) 10 x 10 cm, is wetted with 0.05 M Tris-HCl buffer at pH 8.5 and blotted with dry filter paper to remove excess buffer. The buffer-wetted paper is mounted on a 1 mm thick glass plate placed on the cooling plate of the horizontal electrofcusing apparatus and connected with buffer soaked strips to electrode vessels containing the same buffer as above. A Visking dialysis membrane is placed onto the top of the buffered paper and pressed to to establish good contact. A rectangular strip of 2 mm silicone rubber with a slot in the middle is then layed on the dialysis membrane. The section of interest is removed from the focused gel and distilled water added to make a slurry. The slurry is then placed in the slot of the silicone membrane and electrophoresis conducted at 800 to 1000 V for 40 to 60 min. The dialysis membrane retains the protein, while the carrier ampholyte migrates into the paper electrode wicks. The efficiency of the ampholyte removal is

Table 6. Cont.

	tested by taking the bottom layer of filter paper, drying and staining with 1 % Amido Black 10B in methanol- acetic acid-water (45 : 5 : 25 v/v) followed by destaining in the same solvent.
8. **Elution of proteins**	The gel layer, directly from the focused gel or from the gel electrophoresed to remove ampholyte, is tranferred into a centrifuge tube and suspended in a 1:1.5 v/v ratio of distilled water. The mixture is centrifuged at 35,000g at 2 °C and the supernatant containing the protein collected.

The device for the electrophoretic removal of ampholytes is shown in Fig. 18.

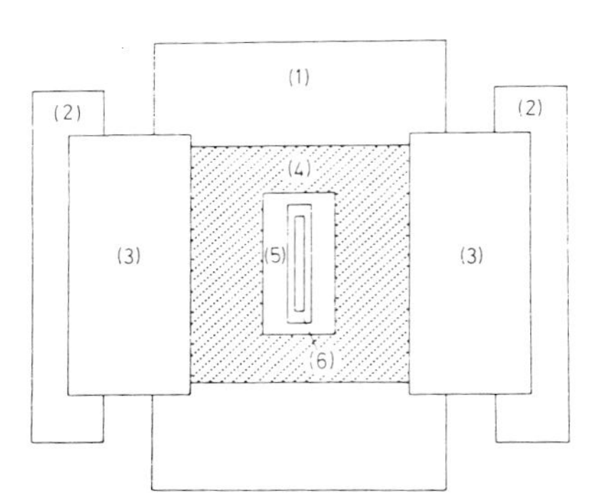

Fig. 18. Plate configuration for removing ampholytes electrophoretically: 1] Cooling plate of the focusing apparatus, 2] electrode vessels, 3] filter paper wicks impregnated with buffer, 4] buffered filter paper on 1 mm glass plate, 5] Visking dialysis membrane, 6] silicone rubber applicator (2mm thick with slot in the middle). (from Frey and Radola) (224)

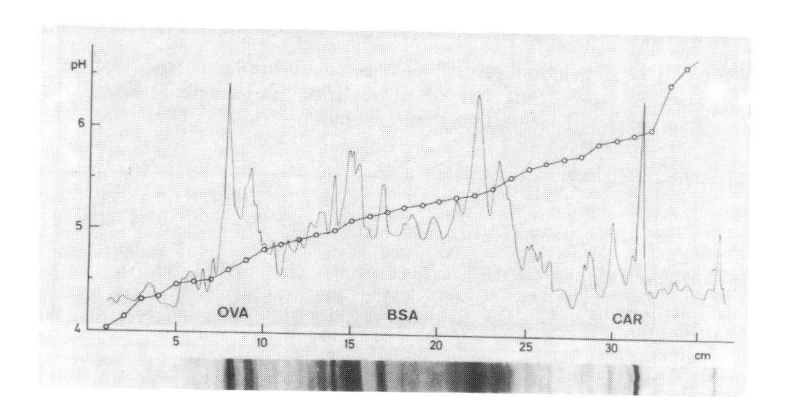

Fig. 19. High resolution preparative IEF of a mixture of ovalbumin (200 mg), bovine serum albumin (200 mg) and carbonic anhydrase (100 mg). Separation distance 40 cm, gel layer 40 x 20 x 0.1 cm, Bio-Gel P-60. Focusing 800 V, 16 h, 3000-5000 V, 5 h; 33,000 Vh. (from Frey and Radola) (224).

3.12. Gel Casting

3.12.1. Gel formulations

The following gel formulations give stock gel reagents to allow one to use either standard isoelectric focusing thin-layer systems or ultrathin-layer techniques discussed in the next section. The examples given are for acrylamide gels with five per cent T and a 3.5 per cent cross linking agent. Other variations may be employed as desired. (See Chapter 2, Tables 7 and 8 pages 52 and 53)

Table 7. Acrylamide stock solution 48 per cent

1. **Acrylamide (recrystallized)**	2.88 gm
2. **BIS acrylamide**	100 mg
3. **Distilled water**	3.30 ml

Warm to bring more rapidly into solution. This amount of stock solution is sufficient for 40 ml of final gel volume. Recrystallization procedures are given in Chapter 1, section 1.1).

Table 8. Standard 5 per cent T, 3.5 per cent C isoelectric focusing gels 100μ to 1mm thick

	1 mm gels	100 - 250μl gels
1. Acrylamide Stock soln.	4.62 ml	1.00 ml
2. **Distilled water**	13.30 ml	3.00 ml
3. **Ampholyte**	2.00 ml	1.00 ml
4. **TEMED**	50 μl	5.0 μl
5. **Ammonium persulfate**	20.0 ml (55 mg/ 100 ml)	5.00 ml (same conc.)

The amount of ampholyte increases from 2 % in the thin-layer gels to 4 % in the ultrathin layer gels. The volumes given will produce one standard 1mm gel 22 cm x 11 cm and five 100μ or three 250μ gels 9 x 10 cm.

3.12.2. Covalent binding of acrylamide gels to glass plates

Gels may be covalently bonded to glass plates treated with agents such as methacryloxypropyltrimethoxysilane (Polyfix 1000™, Desaga GMBH, PO Box 101969, 6900 Heidelberg, West Germany). Acid cleaned glass plates are immersed in a solution of 0.02 per cent Polyfix in 95 per cent ethanol for three to four minutes. They are then allowed to drain for several minutes and are wiped dry with a soft paper towel. Note: It is suggested that rubber gloves be worn so that the silane does not come in contact with the skin. This procedure removes excess Silane from the plate which can later cause background spots, particularly, with silver stain procedures. The dried plates then may be stored for at least several weeks before use for casting gels thereon. Just prior to casting, the plates should be wiped again with a paper towel wetted with 95 per cent ethanol for the same reason. This will in no way interfere with the bonding of the polyacrylamide gel to the plate, but will aid in providing a clearer background following silver staining.

Laquer-treated mylar sheet material may also be treated with silane to bond acrylamide gel, however, this does not always produce as satifactory results as obtained on glass. At present, two commercially available pre-treated mylar films may be better used for this purpose. The first is Gel-Fix™ (Serva Fein Biochemicals) and the second is GelBond-PAG™ (Marine Colloids). Both of these products are suitable for gels up to 0.5mm in thickness. There are two major advantages of the thinner and flexible mylar backings: The first is that, even though the heat transfer characteristic of mylar is poorer than glass, these much thinner backings facilitate better cooling. Second, the flexible mylar backed gels facilitate replicate printing and other transfer techniques given in Chapter 5. Alternatively, acrylamide gel may be bonded directly to GelBond™ if it has been treated previously with 20 per cent glycerol, in order to provide a hydrophilic surface, and the treated sheet dried at 65 °C. This method is not as effective as with the newer materials mentioned above, since occasionally a gel may lift from a glycerol treated mylar sheet.

3.12.3 Casting thin-layer gels

Thin-layer gels may be cast in the vertical position using neoprene or plastic gaskets and bulldog clamps as shown in Fig. 20.

3.12.4 Ultrathin-layer gels

As indicated in Fig. 21, a sandwich of glass and mylar rectangles are used to produce the mold for casting ultrathin-layer gels. The gel thickness is determined by the spacer material used. In the examples shown, rectanglular frames of Parafilm™ are cut to the dimensions of either the silanized glass plate on which the gel is to be backed; or to the size of the mylar rectangle if the gel is to be backed on treated mylar. The Parafilm frame is cut so that at least a 3mm wide boarder is maintained with the outer dimensions matching that of the glass plate or mylar rectangle. Either a new single edge razor blade or scalpel may be used for cutting. The paper backing on the Parafilm should be left in in place for ease of handling and mounting on the plate or mylar. One layer of Parafilm provides a gasket of 125μ, two layers 250μ and so on. When gel thicknesses of over 375μ are desired other casting gasket materials should be employed, such as neoprene or plastic.

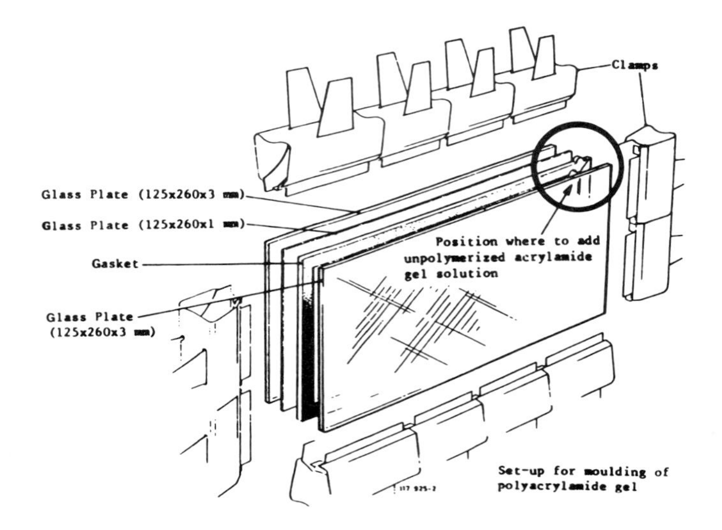

Fig. 20 Schematic setup for casting thin-layer isoelectric focusing gels. (Courtesy of LKB)

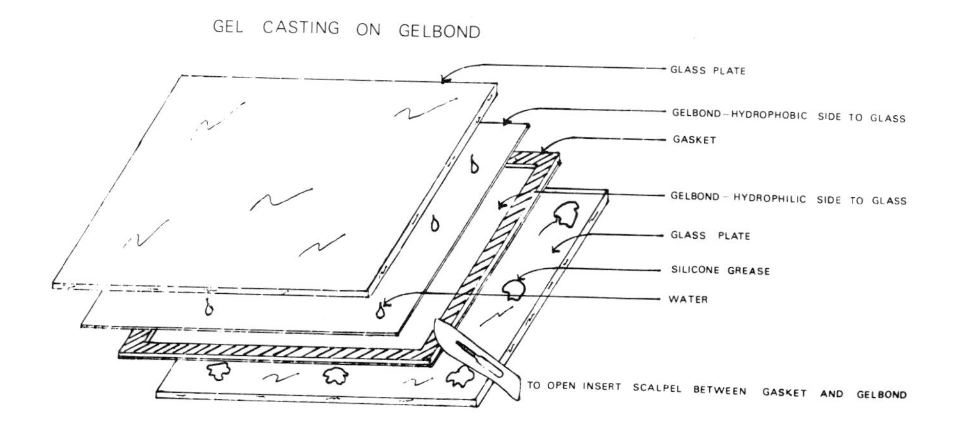

Fig. 21. Casting gels covalently bonded to glass or prepared mylar films. A, glass casting plate; B, Untreated mylar sheet adheared to glass plate with a film of alcohol; C, Parafilm gasket (1-3 layers thick); D, Silane treated glass plate 1mm thick. When bonding on prepared mylars, place the mylar on a glass casting plate with the treated side up and lay the gasket material on this surface. The entire sandwich may be stored at 4 °C in heat sealed bags fabricated from Marvel Seal™, Ludlow Corp. for at least one month.

3.13. High Voltage Focusing

3.13.1. Ultrathin-layer isoelectric focusing

Ultrathin-layer isoelectric focusing is presently accepted to mean focusing on polyacrlamide gels 50 µ to 250 µ thick. Gels 0.5 mm to 1.0 mm thick may be considered to be "so called" thin-layer, which leaves, by present convention, gel thicknesses between 250 and 500 µ rather in limbo as to what they should be designated. Görg *et al.* (225) reported this technique using 120 to 360 μ thick gels backed on cellophane. Radola (226) provided a more practical technique by covalently bonding 50 to 100 μ gels on mylar or thin glass plates. The advantage of ultrathin-layer isoelectric focusing is that higher field strengths, with their increased resolution potential, may be employed. The inherent improved cooling characteristics of ultrathin-layer gels also allow ampholyte concentrations to be increased two-fold or more to improve gradient production and at the same time to reduce potential protein-protein or protein-ampholyte interaction when higher ionic strengths are employed. Thus for example, using 125u thick gels with four per cent ampholytes voltage gradients up to 750V/cm may be employed with appropriate apparatus such as the MRA "Cold Focus". While 125u thick gels provide optimal resolution with protein stains either Coomassie Brilliant Blue R or with the highly sensitive silver stains, enzyme reactions are carried out with less streaking problems, on gels 250u thicker even though there may be some loss of resolution. This is apparently due to the fact that the former produce more surface staining, while the latter allow rapid staining of enzymes within the gel; usually within 10 to 12 min with conventional methods and 1 to 2 min with newer techniques developed by Kinzkofer and Radola (227), which are given in Chapter 5. Longer times may lead to a resolution loss due to diffusion, as demonstrated in Fig. 22.

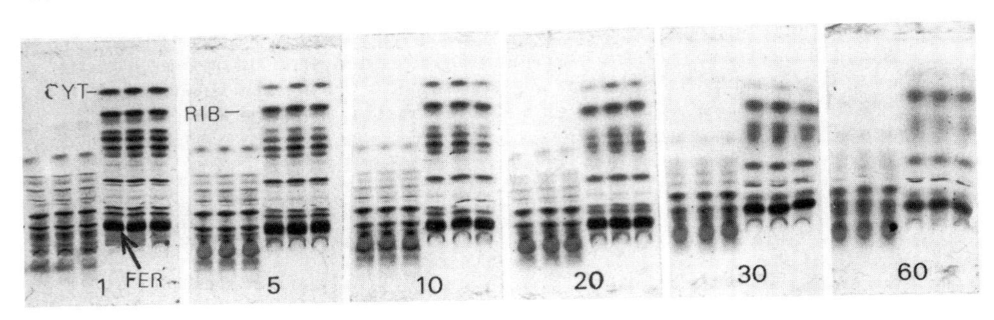

Fig. 22. Diffusion of a mixture of Rohament P and marker proteins CYT-cytochrome c (MW 12,400), RIB-ribonuclease (MW 13,700) and Fer-ferritin (MW 465,000) following standing a 37 °C for 1,5,10,20,30 and 60 min respectively before fixation and staining. Low MW markers spread rapidly while ferritin is little affected. (from Kinzkofer and Radola) (227)

Ampholyte concentrations of four per cent in ultrathin-layer gels do not interfere with subsequent pH equilibration due to the thinness of the gel and rapid diffusion of buffer into the gel. Normally substrate, dye and buffer can be added together without affect on the reaction rate. Extremely high voltages, without specialized cooling, may be applied to 50 to 100 μ gels over very short separation distances, *e.g.*, the 3 cm 50 μ gels described by Radola (228).

To achieve similar voltage gradients and resolution on larger and 125 to 250 μ gel, additional cooling capacity is required. Two companies have developed high efficiency cooling systems using Peltier cooling devices that allow isoelectric focusing to be carried out on gels up to 250μ thick at voltage gradients 2 to 4 times those normally employed. The Cold Focus system (MRA Corp.,Clearwater, FL) and a larger format system from Hoeffer Co. provide similar systems for high voltage isoelectric focusing on ultrathin-layer gels after the methods described by Allen (13). These systems are designed to produce rapid heat dissipation of the Joule heat produced during isoelectric focusing, particularly at high voltage gradients (400-650 V/cm), which produce improved resolution. The beryllium oxide ceramic heat exchange plate provided in the MRA system is 256 times more efficient than glass and some six times more efficient than the metallic oxide ceramic (aluminum oxide) used in the Hoeffer system. The additional cooling capacity, thus achieved, allows greater flexibility in the choice of operating conditions for the improvement of resolution. This is particularly true for ultrathin-layer gel isoelectric focusing where voltage gradients of 400-600 V/cm may be employed, not only to enhance resolution, but also to reduce focusing times to 30 min. or less, depending on the electrode distance and pH gradient. Illustrative given in the following tables may be varied somewhat from the conditions given to suit other ampholytes, or the user's specific requirements.

The rate of temperature decrease in these instruments is dependent on the ambient room temperature for a given amperage input to the Peltier cooling devices. Ambient temperatures above 24 ºC will, for example, limit voltage gradient levels that may be applied to 125μ thick gels to 425-450 V/cm while at 20 ºC 500 plus volts per cm may be employed. Under extreme conditions of ambient temperatures, the apparatus may be operated in a cold room or a refrigerator. However, caution must be exercised under such operating conditions, since the plate temperature at low amperage settings could freeze the gel during the prefocusing period carried out at low voltage. Conditions for various ampholytes and pH ranges on both ultrathin-layer acrylamide gels are shown in the following tables.

Table 9. Cold focus conditions for 125μ thick gels on pH 3-7 gradients. Ambient temperature 22 ºC, separation distance 7 cm

Time	Volts	Watts	Plate	Volt/H
10 min	200	1.0	17 ºC	50
Load samples				
5 min	350	1.0	17 ºC	89
6 min	950	2.0	17 ºC	180
6 min	1000	6.0	17 ºC	350
5 min	2100	12	17 ºC	455
3 min	3000	16	18 ºC	650

Table 10. Cold focus conditions for 250μ thick gels on pH 3-7 Servalyte gradients. Ambient temperature 20 ºC, separation distance 7 cm

Time	Volts	Watts	Plate	Volt/H
10 min	200	1.0	17 ºC	45
Load samples				
5 Min.	320	1.0	17 ºC	84
6 min	550	3.0	17 ºC	162
6 min	1100	6	17 ºC	265
5 min	2000	12	17 ºC	435
2 min	3000	31	21 ºC	550

To attempt to include all recommended focusing conditions for the variety of systems available, is certainly beyond the purpose or scope of this book. The reader is directed to the manual which comes with each apparatus for this information. Presently, at least two companies are supplying greatly detailed texts with their systems: Pharmacia has an excellent "in house" produced book and LKB provides a copy of Righetti's book (2) with their system.

3.14. Agarose Isoelectric Focusing

3.14.1. Casting agarose isoelectric focusing gels

Table 11. Materials and apparatus required

1. IsoGel agarose	15. GelBond film (FMC)
2. Ampholytes	16. d-sorbitol
3. Trisodium EDTA	17. Glycerol
4. Gel casting assembly	18. Distilled water
5. Sodium hydroxide	19. Acetic acid, glacial
6. 20 %trichloroacetic acid	20. Sulfosalicylic acid
7. Methanol	21. Coomassie Brilliant Blue R-250
8. Crocein Scarlet	22. Triton x-100
9. Isoelectric focusing chamber	23. Constant power supply
10. Refrigerated circulator bath	24. Forced-air, heated
11. Refrigerator	25. Hot plate, magnetic stirrer
12. Thermometer (0-120 °C)	26. Soft rubber roller, artist type
13. 20cc syringe (needle 14-16 gauge)	27. 1cc tuberculin syringe
14. Tape, waterproof	28. Parafilm

Fixation and stain procedures using Crowle's double stain for agar focusing are given in Chapter 5, section 5.9.1.

Table 12. Isoelectric Focusing agar gel Preparation

1. **Agarose preparation**	a. Weigh out 1.25 g IsoGel agarose, 8 g d-sorbitol, 3.0 g glycerol and 0.05 g trisodium EDTA. b. Tare a 225 mL Erlenmeyer flask containing a magnetic stirbar.and add approximately 80 mL (80 g) distilled water to the flask. Then, slowly add the agarose to the room temperature water with continuous stirring, allow to disperse, and add the sorbitol, EDTA and 3 mL of glycerol with stirring. c. Add distilled water to 100 g and (weigh flask and contents). Place an inverted Erlenmeyer of smaller size in the mouth of the flask and place in boiling water bath. Still continually stirring, heat the contents to boiling and maintain gentle boiling for 10 minutes (At the same time, heat a beaker of distilled water to boiling for later use to bring the boiling gel to the proper volume and to warm a syringe for transfer of the liquid agarose. d. Weigh the flask and contents (minus the inverted flask)and add warm water to the original weight and then aliquot the warm agarose into glass tubes (*e.g.* 10 mL), cover with Parafilm, label, and store at 4 ºC (The tubes can be stored several weeks at 4 ºC).
2. **Casting the agarose gel**	a. Take the agarose-containing tube (*e.g.* 10mL) from the refrigerator, remove the Parafilm and place an inverted beaker over the tube mouth and place the tube in a boiling water bath. While heating is underway, assemble the casting cassette. (Fig. 23). This consists of: 1a. Spreading several ml of 0.1% Triton x-100 solution on the glass plate. 2a. Placing a piece of GelBond film of the same dimensions hydrophilic side up on the glass plate excluding air bubbles between the glass and GelBond and roll flat with an artists' roller as shown in Fig.24. Carefuly wipe the excess fluid at the edges, Place the plastic cover over the GelBond and glass plate. Clamp the assembly together with large bulldog clips. 3a. Warm the casting assembly to 60 ºC in an oven. Avoid overheating since this will cause the GelBond to buckle off the glass and an uneven gel will be formed. 4a. Once the agarose in the tube has completely dissolved, place it in a dry heating block at 60-65 ºC and allow the agarose to come to this temperature.

Table 12. Cont.

	b. In a separate glass tube, warm ampholytes (sufficient to give a 2.5% solution (*e. g.* 0.63 mL) in the dry heating block. Two minutes before removing the casting assembly from the oven pour the agarose into the ampholyte tube, cover the tube with several layers of Parafilm and slowly invert the tube several times to throughly mix the ampholytes and agarose with minimal aeration. Allow the agarose-ampholyte mixture to sit several minutes in the dry heating block to allow air bubbles to rise to the surface. c. Flush the 20 mL syringe with boiling water a few times to throughly heat it. Expel all water from the barrel and needle and immediately fill the syringe with the warm agarose mixture and inject the mixture into the opening of the prewarmed casting assembly. It is important to use small pulsating pushes of agarose to flow to the bottom of the casting assembly excluding air as it fills the assembly. d. Seal the opening of the casting assembly with tape to prevent evaporation. Allow the casting assembly to cool to room temperature, then tranfer it to a 4 ºC refrigerator for at least 20 min. Remove the slab gel plate from the casting assembly, place in a humidity chamber, seal and refrigerate for at least 1 h.
3. **Preparation for electrofocusing**	a. Turn on the refrigerated circulator bath and allow to equilibrate at 10-15 ºC before proceeding. b. Take the agarose slab gel-containing humidity chamber out of the refrigerator and allow the gel to equilibrate at room temperature for 30 minutes and trim the edges of the slab gel to a desired size. c. Place several mL of 0.1% Triton x-100 on the cooling platen of the electrophoresis chamber and carefully lay the GelBond-supported agarose slab gel onto the Triton x-100 and platen making sure that there is uniform contact between the film support and platen with no entrapment of air bubbles Wipe the edges around the film support several times to make sure there is no excess fluid. d. If surface fluid is present, blot the surface of the gel briefly with a sheet of hard surface filter paper such as Schleicher and Schuell #577 in order to remove excess fluid from the gel surface. It has been found advantageous to overlay the gel surface with a piece of hydrophobic plastic film for several minutes. This brings the fluid contents of the gel to equilibrium in all parts of the gel.

Table 12. Cont.

	e. Place the Mylar applicator mask onto the gel and press the mask carefully on the gel, avoiding air spaces between the mask and the wells (The applicator mask is a 4 mil thick piece of Mylar with openings cut into it to receive the sample. f. Apply 2-5 μl samples into the applicator wells. (The sample should be at a 2-6 mg/mL concentration.) Samples can include tissue sections or cells with or without nonionic detergent such as Chaps buffer or Triton X-100. g. Cut two electrode wicks (Schleicher – Schuell #470) to the length of the gel. Soak one in the anolyte and the other in the catholyte. The anolyte-soaked wick should be quite wet, but not dripping, for best results. The catholyte wick should be blotted lightly for best results. Place the wicks on the gel surface so that they lay parallel to each other at the ends of the gel, and extend half over the gel ends. Make sure that the electrodes are clean and that there are no breaks in the wire or ribbon electrodes. Place the electrodes onto the wicks on that part which is off the gel, making sure that they are parallel. Inspect the apparatus to make sure all electrical contacts are clean and secure and close the cover of the apparatus before turning on the power supply.

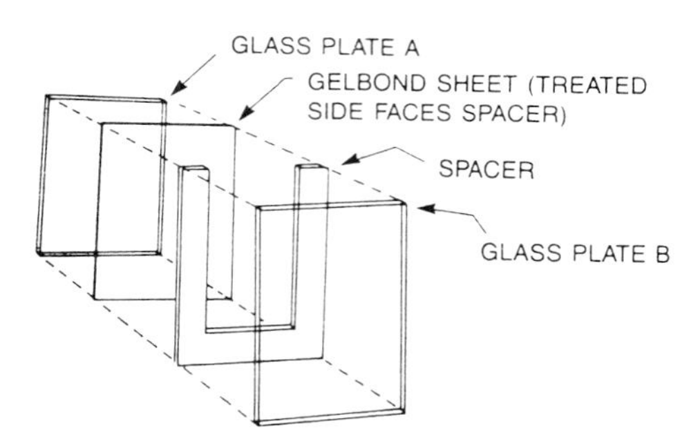

Fig. 23. Cassette for casting agarose gels. (from Cook and Saravis) (229)

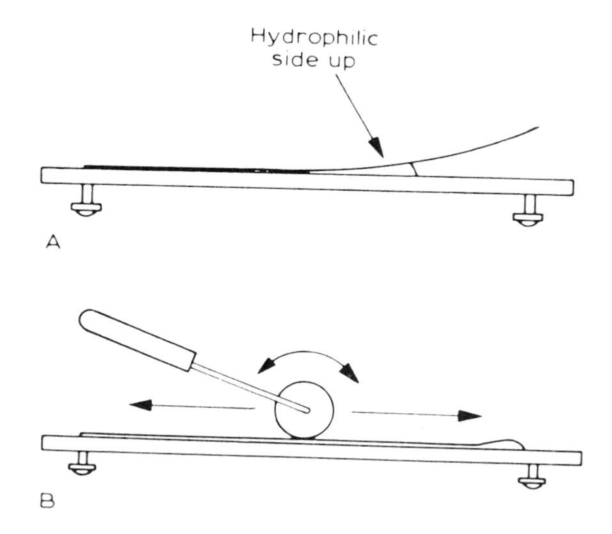

Fig. 24. Procedure for rolling GelBond onto glass plate wetted with Triton X 100. (from Cook and Saravis) (229)

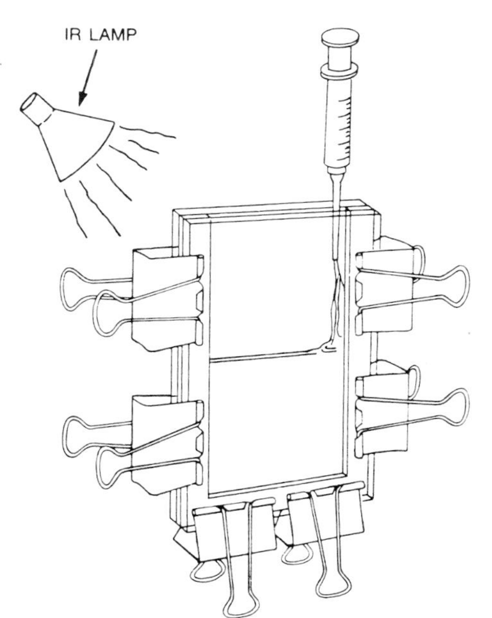

Fig. 25. Schematic diagram of casting hot agarose in a cassette. (from Cook and Saravis) (229).

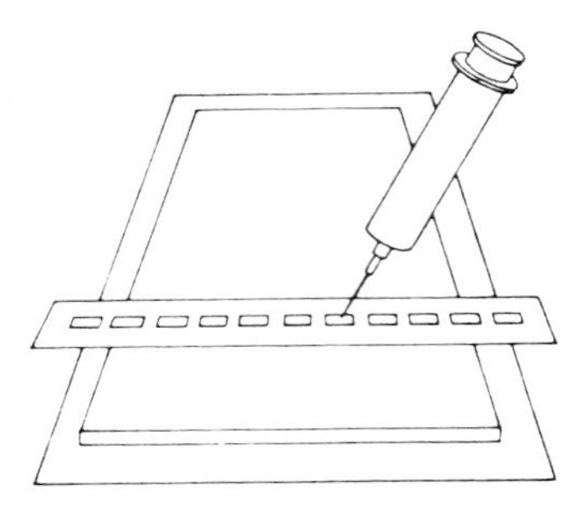

Fig. 26. Application of samples into wells cut in a Mylar film strip. This method is equally applicable to PAGIF

Table 13. Run conditions for agarose IEF on 0.8mm gels at a variety of pH ranges

pH range	3.5 - 9.5	2.5 - 4.5	4.6 - 6.5	5.0 - 8.0
Voltage (limit)	1500	250	500	500
Power	25 W	10 W	10 W	10 W
Anolyte	0.5 M Acetic acid	1.0 M H_3PO_4	2 % pH 2.5-4 ampholyte	2 % pH 2.5-4 ampholyte
Catholyte	1.0M NaOH	0.05 M Threonine	0.1 M Bicine	0.1 M NaOH
Run time	30 min	90 min	90 min	90 min

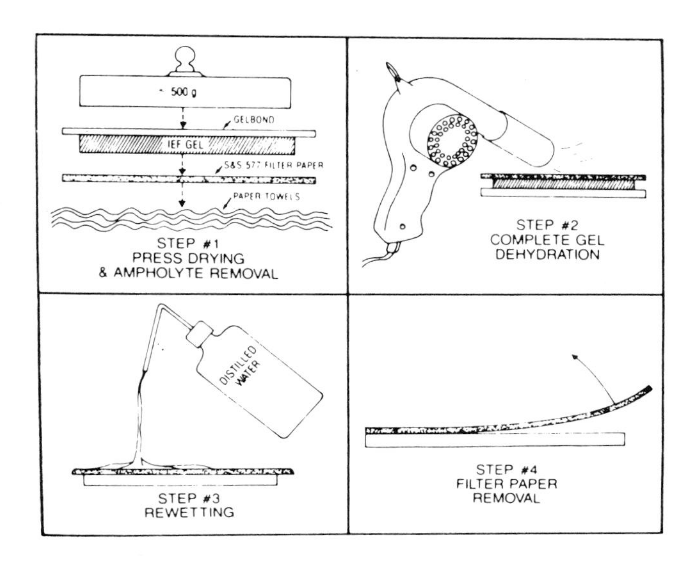

Fig. 27. Procedure for drying an agarose IEF gel prior to staining. After fixing in TCA, the gel is presses for 30 min against filter paper and paper towels with a weight of 0.5 to 1 kg. The dehydration is completed with a hair dryer (step 2)and then the filter paper is removed from the agarose film by a quick rewetting step (3 and 4). (from Cook and Saravis) (229)

Good electrofocusing results in agarose gels are dependent on many factors, a number of which have been addressed above. An example of this technique is shown in Fig. 28. The staining procedure using Crowle's double stain technique is given later in Chapter 5, page 233.

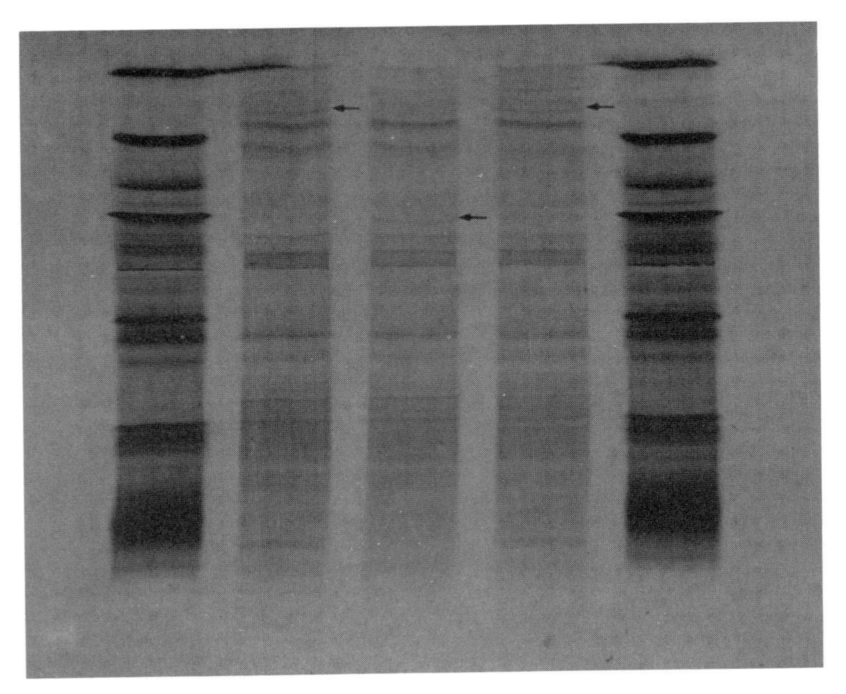

Fig. 28. Direct tissue isoelectric focusing carried out on 0.8mm agarose. From left to right: pI markers, undifferentiated human tumor tissue culture cells, well differentiated tumor tissue culture cells, poorly differentiated tumor tissue cells, pI markers. 150,000 cells were run on each application. (from Saravis)

Using these techniques and information, it is hoped that the individual investigator will be able more readily to approach each separation problem and achieve optimum running conditions without having to resort to long and tedious trial and error methods. Standardized technique between laboratories still remain a hope for the future, but it is obvious that the higher the voltage gradient the better the resolution. As in PAGIF, limiting conditions for resolution are: the voltage potential applied across the gel, heat tranfer capacity of the apparatus, ampholytes used, and concentration of ampholytes.

3.15. Gradient Control

The above procedures and methodology are contingent also on the control of the pH gradient. In broad range pH gels, the commonly used 1.0 **M** phosphoric acid anolyte and 1.0 **M** sodium hydroxide catholyte, work well in PAGIF. In agarose focusing, the anolytes and catholytes in the last section cover the majority of applications. In focusing with gels 1mm thick or over, attention must be paid to maintaining an excess of electrolyte throughout the run. This is particularly true in buffer focusing, where the anolyte is rapidly expended. In ultrathin-layer focusing, the volume of the gel is relatively small in comparison to that of the buffer wicks and few problems arise from using the common H_3PO_4 and NaOH. As a rule, both the anolyte and catholyte should bracket the two extremes of the pH gradient selected. Therefore a number of the common anolytes and catholytes are given in Table 14.

Table 14. Electrolytes for isoelectric focusing

Anolyte	Concentration	pH (25 °C)
Phosphoric acid	1.0 M	1.0
Sulfuric acid	0.2 M	1.6
Acetic acid	0.5 M	2.6
L-glutamic acid	0.04 M	3.2
Indole acetic acid	0.003 M	3.8
L-tyrosine	0.004 M	4.5
Catholyte		
Threonine	0.05 M	5.8
Glycine	0.05 M	6.15
Hepes	0.4 M	7.3
Histidine	0.04 M	7.35
Bicine	0.1 M	8.0
Sodium hydroxide	1.0 M	13

3.15.1. Gradient broadening

Gradients may be broadened to increase resolution of closely spaced bands. An example is the inclusion of β-alanine, (pI) 6.9, as a spacer to increase the separation distance between hemoglobin A_{1c} and HbA_{11}, thus; increasing the reliability of it measurement in normal and diabetic patients (230). This procedure can be used with a variety of amino acid and buffer spacers. The recent introduction of Immobilines, with their potential of very narrow pH ranges and high resolution, should provide a better means for this purpose in the future.

3.15.2 Gradient flattening

An additional method to effect gradient broadening without the introduction of spacers has been described by Altland and Kaempfer (231). Flattening and shifting of the pH gradient, in this case is accomplished by local increases in the gel volume. An example of this technique is shown in Fig. 29.

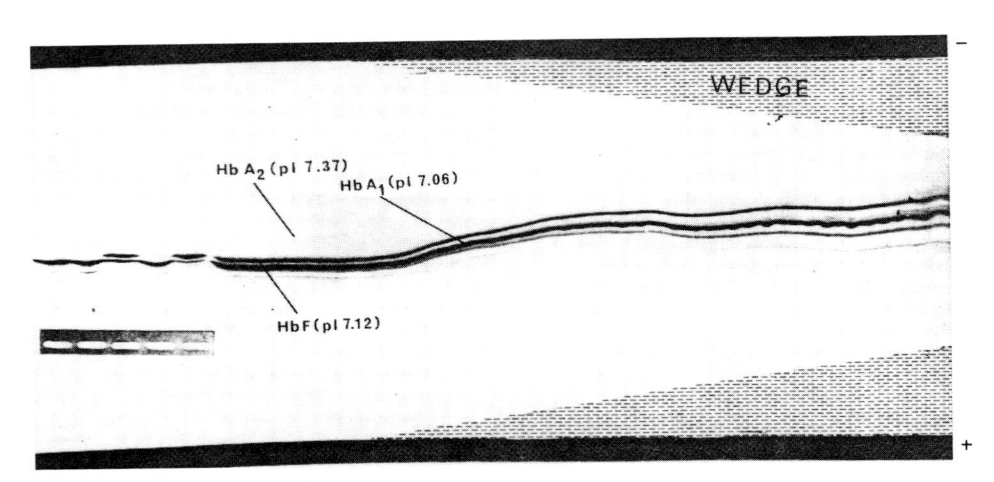

Fig. 29. Continuous flattening of the pH gradient to increase the separation of hemoglobins. The base gel was 0.35mm thick containing 2 % pH 6 to 8 Servalyte, while the gel wedges at anode and cathode were 2mm thick and contained the same ampholyte. (from Altland and Kaempfer) (231)

3.16. Trouble Shooting Guide

Table 15. Common problems and solutions in isoelectric focusing

Problem	Cause	Solution
1. **Streaks or gaps**	A. Particulates present	A. Centrifuge sample
	B. Isoelectric precipitation	B. Remove sample deposit from gel surface with a tissue before removing sample application mask.
		B1. Applying sample in a different location
3. **Fuzzy bands**	A. Not focused long enough	A. Increase focusing time or increase voltage at end of run for a few minutes
	B. Focusing too long	B. Decrease focusing time by incrementally increasing voltage gradient
4. **Skewed bands**	A. Electrode not clean	A. Clean electrode
	B. Uneven electrode contact	B. Level instrument and add weight to maintain good electrode contact.
	C. Sample applied too close to the edge of the gel	C. Do not apply sample closer than 10 mm from the edge of the gel
	D. Electrode wicks too short	D. Wicks should extend to the edge of the gel
5. **Missing or faint bands**	A. Low MM proteins (< 10 Kd)	A. Immunofixation or glutaraldehyde fixation
	B. Protein not denatured by fixative	B. Increase TCA conc., prefix with Glutaraldehyde)
6. **Wavy bands**	A. Salt in sample	A. Desalt sample
	B. Improper anolyte	B. Replace anolyte

Table 15. Cont.

Wavy bands cont.	C. Old electrolyte solution	C. Make fresh
	D. Improper voltage	D. Initially reduce to 30 V/cm
	E. Dirty electrodes	E. Clean thoroughly
6A. Wavy bands at high voltage gradients only	A. Old acrylamide	A. Recrystallize
7. Arc-shaped bands	A. Thin spot in gel	A. Reduce evaporation by keeping gel sealed until used
8. Burning or sparking	A. Dry wicks	A. Anodal wick should be wet; cathodal wick should be damp
	B. Conductivity gap in ampholytes	B. Supplement deficient region. Use mixture from two manufacturers, or increase ampholyte concentration and use ultrathin-layer
	C. Focusing too long	C. Reduce focusing time, incementally increase voltage gradient
	D.Excess salt in sample	D. Dialyze
9. Sample smearing	A. Sample overload	A. Decrease sample concentration
	B. Sample aggregation	B. Solubilize sample with detergent or 2% ampholyte
10. Fluid expression	A. At cathode only	A. Blot cathodal wick with filter paper until damp
	B. Localized elsewhere	B. Salt in sample - desalt
	C. Sample overload	C. Decrease sample concentration
	D. Uneven electrode contact	D. Weight electrode wicks
	E. In Agarose- concentration too low (< 1%)	E. Increase agarose concentration to 1 or 1.25%

Table 15. Cont.

Fluid expression cont.		
	E1. Inadequate curing of Agarose	E1. Cure 1 hour under refrigeration
	E2. Agarose gel was not adequately blotted	E2. Use blotting paper
	E3. Absence of humectants	E3. Add 10% (W/V) sorbitol and 5% (W/V) glycerol
11. Condensation inside chamber	A. Inefficient cooling	A. Adjust cooling temperature to 10 °C or lower
	B. Too high voltage setting	B. Adjust to 1W for 1st 10 min, voltage input should not increase gel temp. more than 25 °C
	C. Humid lab conditions	C. Add air conditioner to lab
12. Incomplete sample uptake	A. Gel too thin	A. Increase gel thickness
	B. In Agarose- sample application mask not left on long enough	B. Allow at least 10 min at 20 V/cm to ensure complete uptake
	B1. Agarose gel too wet when sample applied	B1. Blot gel before sample application
	C. Aggregates in sample	C. Centrifuge sample
		C2. Remove sample deposit from gel surface with a cotton tipped applicator
13. Gel Thinning	A. At electrode	A. Improper electrolyte
	B. Near point of application	B. Sample overloaded
	C. Gel stored too long under low relative humidity	C. Place wetted paper towel in storage bag and seal well
14. Uneven pH gradient	A. Electrodes not making parallel contact	A. Level plate and electrodes

Table 15. Cont.

Uneven pH gradient cont.		
	B. Ampholyte concentration too low	B. Ampholyte concentration should be increased
	C. Impure gel	C. Acrylamide - recrystallize, agarose - try new lot
15. **Loss of basic region of pH gradient**	A. Cathodic drift	A. Supplement gel with pH 9-11
		A1. Decrease focusing time
16. **Current increases with time**	A. Electrodes reversed	A. Throw gel out and have a cup of coffee
17. **Agarose peeling off GelBond backing**	A. Insufficient gel curing time	A. Cure gel for 1h at 4-8 °C
	B. Gel was cast on the wrong (*i.e.* hydrophobic) side of GelBond	B. Test GelBond surface with water droplet before casting. Droplet spreads on hydrophilic side
17A. **Acrylamide peeling**	A. Glass - silane no good	A. Use fresh silane
18. **High background stain**	A. Residual ampholytes	A. Agar follow recommended press-blotting
		A1. Acrylamide - insufficient fixation time
	B. Old stain	B. Make up fresh stain assure all dye is in solution
	C. Gels allowed to settle to bottom of staining dish	C. Use shaking water bath
19. **Diagonal band migration**	A. Uneven electrode contact	A. Apply electrodes evenly on wicks, level instrument and cooling plate
	C. Uneven Electrode positioning	C. Ensure that electrodes are parallel

3.17. References

1. Righetti, P. G. and Drysdale, J. W. : *Isoelectric Focusing*, North Holland, Amsterdam, (1983).

2. Righetti, P. G. : *Isoelectric Focusing: Theory, Methodology and Applications*, Elsevier Biomedical Press, Amsterdam (1983).

3. Righetti, P. G., Van Oss, C.J. and Vanderhoff, J. (eds.) : *Electrokinetic Separation Methods*, Elsevier, Amsterdam (1976).

4. Righetti, P. G. : *Separ. Purif. Methods*, *4*, 23 (1975).

5. Rhigetti, P, G., Gianazza, E. and Bianchi Bosisio, A. in Frigerio, A. and Renoz, L. (eds.), : *Recent Developments in Chromatography and Electrophoresis*, Elsevier, Amsterdam, p. 1 (1979.

6. Bishop, R. : *Sci. Tools*, *26*, 4 (1979).

7 Chrambach, A. : *Mol. Cell Biochem.* *29*, 23 (1980).

8. Righetti, P. G., Gianazza, E. and Bianchi Bosisio, A. in Firgerio, A. and McCamish, M. (eds.), : *Recent Developments in Chromatography and Electrophoresis 10*, Elsevier, Amsterdam, p. 89 (1980).

9. Allen, R. C. and Arnaud, P. (eds.), : *Electrophoresis '81*, de Gruyter, Berlin (1981).

10. Stathakos, D. (ed.), : *Electrophoresis '82*, de Gruyter, Berlin, (1983).

11. Hirai, H. (ed.), : *Electrophoresis '83*, de Gruyter, Berlin, in press.

12. Hjelmeland, L. M. and Chrambach, A. : *Electrophoresis 4*, 20 (1983).

13. Allen, R. C. : *Electrophoresis 1*, 32 (1980).

14. Bjellqvist, B., Ek, K., Righetti, P. G., Gianazza, E., Görg, A. and Westermeier, R. : *J. Biochem. Biophys. Methods*, *6*, 317 (1982).

15. Kolin, A. : *Nat. Acad. Sci. USA*, *41*, 101 (1955).

16. Svensson, H. : *Prot. Biol. Fluids*, *15*, 515.

17. Svensson, H. : *Acta Chem. Scand.* *15*, 325 (1961).

18. Svensson, H. : *Acta Chem. Scand.* *16*, 456 (1962).

19. Vesterberg, O. : *Acta Chem. Scand.* *23*, 2653 (1969).

20. Pogacar, P. and Jarecki, R. : in Allen, R. C. and Maurer, H. R. (eds.), *Electrophoresis and Isoelectric Focusing in Polyacrylamide Gels*, de Gruyter, Berlin, p. 153 (1974).

21. Grubhofer, N. and Borjia, C. : in Radola, B. J. and Graesslin, D. (eds.), : *Electrofocusing and Isotachophoresis*, de Gruyter, Berlin, p. 481 (1977).

22. Williams, K. W. and Soderberg, L. : *Am. Lab. 11*, 97 (1979).

23. Kolin, A. : *J. Chem. Phys. 22*, 1628 (1954).

24. Strangin, A., Levin, E. D. and Stepanou, V. M. : *Biochimiia, 38*, 1098 (1973).

25. Peterson, E. : *Acta Chem. Scand. 23*, 261 (1969).

26. Johnsson, M. : *Acta Chem. Scand. 26*, 3435 (1972).

27. Nguyen, N. Y. and Chrambach, A. : *Electrophoresis 1*, 14 (1980).

28. Prestridge, R. L. and Hearn, M. T. W. : *Anal. Biochem. 98*, 95 (1979).

29. Cuono, C. B. and Chapo, G. A. : *Electrophoresis 3*, 65 (1982).

30. Cuono, C. B. : in Hirai, H. (ed.), *Electrophoresis '83*, de Gruyter, Berlin, in press.

31. Cuono, C. personal communication.

32. Fawcett, J. S. personal communication.

33. Rilbe, H. : *Ann. N.Y. Acad. Sci. 209*, 11 (1973).

34. Giddings, J. C. and Dahlgren, H. : *Sep. Sci. 6*, 345 (1971).

35. Righetti, P. G. and Drysdale, J. W. : *Biochem. Biophys. Acta 236*, 17 (1971).

36. Chrambach, A. Doerr, P., Finlayson, G. R., Miles, L. E. M., Shering, R. and Rodbard, D. : *Ann. N.Y. Acad. Sci. 209*, 44 (1973).

37. Fawcett, J. S. : in Righetti, P. G. (ed.), *Isoelectric Focusing and Isotachophoresis*, North Holland/American Elsevier, p. 25 (1978).

38. Righetti, P. G. : in Righetti, P. G. Van Oss, C. J. and Vanderhoff, J. W. (eds.), *Electrokinetic Separation Methods*, Elsevier North Holland, Amsterdam, 389 (1979).

39. Gianazza, E., Astorri, C. and Righetti, P. G. : *J. Chromatogr. 171*, 161 (1979).

40. Görg, A., Postel, W. and Westermeier, R. : *Anal. Biochem. 89*, 60 (1978).

41. Radola, B. J. : in *Electrophoresis '79*, Radola, B. J. (ed.), de Gruyter, Berlin, p. 79 (1980).

42. Allen, R. C., Oulla, P. M., Arnaud, P., and Baumstark, J. S. : in *Electrofocusing and Isotachophoresis*, Radola, B. J. and Graesslin, D. (eds.), de Gruyter, Berlin, p. 255 (1977).

43. Allen, R. C. and Arnaud, P. : *Electrophoresis 3*, 205 (1983).

44. O'Farrell, R. : *J. Biol. Chem. 250*, 4007 (1975).

45. Anderson, N. G. and Anderson, N. L. : *Anal. Biochem. 85*, 331 (1978).

46. Anderson, N. L. and Anderson, N. G. : *Anal. Biochem. 85*, 341 (1978).

47. Hjelmeland, L. M., Nebert, D., W. and Chrambach, A. : in Catsimpoolas, N. (ed.), *Electrophoresis '78*, Elsevier, North Holland, N.Y. p. 29 (1978).

48. Hjelmland, L. M. and Chrambach, A. : *Electrophoresis 2*, 1 (1981).

49. Hjelmeland, L. M. and Chrambach, A. : in Jacoby, W. B. (ed.) *Methods in Enzymology, 104*, Academic Press, New York p. 105 (1983).

50. Makino, S. : *Adv. Biophys. 12*, 131 (1979).

51. Lu, A. Y. H. and Levin, W. : *Biochim. Biophys. Acta 344*, 205 (1974).

52. Tanford, C.: *The Hydrophobic Effect*, 2nd Edition, J. Wiley, New York, (1980).

53. Hjelmeland, L. M. : *Proc. Nat'l Acad. Sci.* USA, *77*, 6368 (1980).

54. Simonds, W. F., Koski, G., Streaty, R. A., Hjelmeland, L. M. and Klee, W. A. : *Proc. Nat'l Acad. Sci.* USA, : *77*, 4623 (1980).

55. Malpartida, F. and Serrano, R. : *FEBS Lett. 111*, 69 (1980).

56. Vyvoda, O. S., Coleman, R. and Holdsworth, G. : *Biochim. Biophys. Acta 465*, 68 (1977).

57. Coleman, R., Holdsworth, G. and Finean, J. B. : *Biochim. Biophys. Acta 436*, 38 (1976).

58. Gianazza, E., Astorri, C. and Righetti, P. G. : *J. Chromatogr. 171*, 161 (1979).

59. Altland, K. : *Clin. Chem. 28*, 1000 (1982).

60. Righetti, P. G., Krishnamoorthy, R., Lapoumeroulie, C. and Labie, D. : *J. Chromatogr. 177*, 219 (1979).

61. Allen, R. C. : *J. Chromatogr. 146*, 1 (1978).

62. Spicer, K. M., Allen, R. C. and Buse, M. D. : *Diabetes, 27*, 384 (1978).

63. Altland, K. : in Radola, B. J. and Graesslin, D. *Electrofocusing and Isotachophoresis*, de Gruyter, Berlin, p. 295 (1977).

64. Maurer, H. R. and Allen, R. C. : *Clin. Chim. Acta 40*, 359 (1972).

65. Rodbard, D. and Chrambach, A. : *Anal. Biochem. 40*, 95 (1971).

66. Watkins, J. E. and Miller, R. A. : *Anal. Biochem. 34*, 424 (1970).

67. Righetti, P. G. : in Allen, R. C. and Arnaud, P. (eds.), *Electrophoresis '81*, de Gruyter, Berlin, p. 3 (1981).

68. Latner, A. L. : *Adv. Clin. Chem. 17*, 193 (1975).

69. Arnaud, P., Wilson, G. B., Koistinen, J. and Fudenberg, H. H. : *J. Immunol. Methods, 16*, 221 (1977).

70. Chapuis-Cellier, C., Francina, A. and Arnaud, P. : in Radola, B. J. (ed.), *Electrophoresis '79*, de Gruyter, Berlin, p. 711 (1980).

71. Lasne, Y., Lasne, F., Benzerra, O. and Arnaud, P. : in Radola, B. J. (ed.), *Electrophoresis '79*, de Gruyter, Berlin, p. 727 (1980).

72. Constans, J., Viau, C., Gouaillard, C., Bouissou, C. and Clerc, A. : in Radola, B. J. (ed.), *Electrophoresis '79*, de Gruyter, Berlin, p. 701 (1980).

73. Altland, K., Roeder, T., Jakin, H. M., Zimmer, H. G. and Neuhoff, V. : *Clin. Chem. 24*, 1000 (1982).

74. Allen, R. C., Arnaud, P. and Spicer, S. S. : in Allen, R. C. and Arnaud, P. (eds.), *Electrophoresis '81*, de Gruyter, Berlin, 167 (1981).

75. Narayanan, K. R. and Raj, A. A. : in Radola, B. J. and Grasesslin, P. (eds.), *Electrofocusing and Isotachophoresis*, de Gruyter, Berlin, p. 221 (1977).

76. Pretch, W., Charles, D. J. and Narayanan, K. : *Electrophoresis 3*, 142 (1982).

77. Williams, A. R., Salaman, M. R. and Kreth, H. R. : *Ann. N.Y. Acad. Sci. 209*, 210 (1973).

78. Awdeh, Z. L., Williamson, A. R. and Askonas, B. A. : *Nature* (London), : *219*, 66 (1968).

79. Kreth, H. W. and Williamson, A. R. : *Protides Biol. Fluids* Proc. : *20*, 189 (1973).

80. Keck, K., Grossberg, A. L. and Pressman, D. : *Eur. J. Immunol. 3*, 99 (1975).

81. Keck, K., Grossberg, A. L. and Pressman, D. : *Immunochem. 10*, 331 (1973).

82. Cornell, F. N. : *Pathology, 7* , 258 (1975).

83. Awdeh, Z. L., Williamson, A. R. and Askonas, B. A. : *Biochem. J. 116*, 241 (1970).

84. Brendel, S., Mulder, J. and Verhaar, M.A.T. : *Clin Chim Acta 54*, 243 (1974).

85. Trieshman, H. W., Abraham, G. N. and Santucci, E. A. : *J. Immunol. 144*, 176 (1975).

86. Dale, G., Latner, A. L. and Muckle, T. J. : *J. Clin. Pathol. 23*, 35 (1970).

87. Cwynarski, M. T., Watkins, J. and Johnson, P. M. : in Arburthnott, J. P. and Beeley, J. A. (eds.), *Isoelectric Focusing*, Butterworths, London, p. 313 (1975).

88. Bouman, H., Meincke, G. and Hausteen, B. : *Z. Immunol.-Forsch. 150*, 370 (1975).

89. Hobart, M. J. and Lachmann, P. J. : *J. Immunol. 116*, 1736 (1974).

90. Hovanessian, A. G. and Awdeh, Z. L. : *Eur. J. Biochem. 68*, 333 (1976).

91. Gaffney, P. J. : *Nature-New Biol. 230*, 54 (1971).

92. Soria, J., Soria, G., Duvand, G. and Feger, J. : *Ann. Biol. Clin.* (Paris), *30*, 61 (1972).

93. Arnesen, H. : *Thrombosis Res. 4*, 861 (1974).

94. Godolphin, W. J. and Stinson, R. A. : *Clin. Chim. Acta 56*, 97 (1974).

95. Utermann, G., Hess, M. and Vogelberg, K. H. : in Radola, B. J. and Graesslin, D. (eds.), *Electrofocusing and Isotachophoresis*, de Gruyter, Berlin, 281 (1977).

96. Jones, S. M., Creeth, J. M. and Kekwick, R. A. : *Biochem. J. 127*, 187 (1972).

97. Frenoy, J. P. and Bourrillon, R. : *Biochim. Biophys. Acta 371*, 168 (1974).

98. Ohlson, K. and Skude, G. : *Clin. Chim. Acta 66*, 1 (1976).

99. Hamberg, U., Turpeinen, U. and Knuutinen, U. : in Radola, B. J. and Graesslin, D. (eds.), *Electrofocusing and Isotachophoresis*, de Gruyter, Berlin, p. 351 (1977).

100. Laurell, C. B. and Eriksson, S. : *Scand. J. Lab. invest. 15*, 132 (1963).

101. Sharp, H. L., Bridges, R. A., Krivit, L. and Freier, E. F. : *J. Lab. Clin. Med. 73*, 934 (1969).

102. Cox, D. W. and Huber, O. : *Lancet, 90*, 1216 (1976).

103. Arnaud, P., Chapuis-Cellier, C., Souillet, G., Carron, R., Wilson, G. B. and Fundenberg, H. H. : *Trans. Ass. Amer. Physiol. 89*, 205 (1976).

104. Allen, R. C., Javed, T., Arnaud, P., Myers, M. and Malena, D. E. : *J. Dent. Res. 58*, 220 (1979).

105. Allen, R. C., Harley, R. A. and Talamo, R. C. : *Amer. J. Clin. Pathol. 62*, 732 (1974).

106. Arnaud, P., Chapuis-Cellier, C. and Creyssel, R. : *Protides Biol. Fluids* Proc. Collog. *22*, 515 (1975).

107. Fagerhol, M. K. and Laurell, C. B. : *Clin. Chim. Acta 16*, 199 (1967).

108. Allen, R. C. and Hennigar, G. R. : *Amer. J. Clin. Pathol. 67*, 209 (1977).

109. Kostner, G., Holasek, A., Schoenburn, W. and Fuhrmann, W. : *Clin. Chim. Acta 38*, 155 (1972).

110. Kostner, G., Albert, W. and Holasek, A. : *Hoppe Seyler's Z. Physiol. Chem. 350*, 1347 (1969).

111. Sodhi, H. S., Sundaram, G. S. and Mackenzie, S. L. : *J. Lab. Clin. invest. 33*, 71 (1974).

112. Blaton, V., Vercaenst, R., Vandercasteele, N., Caster, H. and Peeters, H. : *Biochemistry, 13*, 1127 (1974).

113. Eggena, P., Tivol, W. and Aladsem, F. : *Biochem. Med. 6*, 184 (1972).

114. Burnstein, M. and Scholnick, H. R. : *Adv. Lipid Res. 11*, 67 (1973).

115. Scanu, A. M., Edelstein, C. and Aggerbeck, L. : *Ann. N.Y. Acad. Sci. 209*, 311 (1973).

116. Zannis, V. I. and Breslow, J.L. in Radola, B. J. : *Electrophoresis '79*, de Gruyter, Berlin, p. 437 (1980).

117. Marshall, J. S., Pensky, J. and Williams, S. : *Arch Biochem. Biophys. 156*, 456 (1973).

118. Alpert, E., Drysdale, J. W. and Isselbacher, : *Ann. N.Y. Acad. Sci. 209*, 387 (1973).

119. Sokolov, A. V., Mekler, L. B., Tsvetkov, V. S. and Gusev, A. I. : *Byull. Eksp. Biol. Med. 75*, 31 (1973).

120. Park, C. M. : *Ann. N.Y. Acad. Sci. 209*, 237 (1973).

121. Bunn, H. F. : *Ann. N.Y. Acad. Sci. 209*, 345 (1973).

122. Drysdale, J. Righetti, P. G. and Bunn, H. F. : *Biochim. Biophys. Acta 229*, 42 (1971).

123. Dozy, A. M. and Huisman, T. H. J. : *J. Chromatogr. 40*, 62 (1969).

124. Huisman, T. H. J. and Meyering, C. A. : Clin. Chim. Acta 5, 103 (1960).

125. Jeppsson, O. and Bergland, S. : *Clin. Chim. Acta 40*, 153 (1972).

126. Taketa, F., Huang, Y. P., Libnoch, J. A. and Dessel, B. H. : *Biochim. Biophys. Acta 400*, 348 (1975).

127. Monte, M., Beuzard, Y. and Rosa, J. : *Amer. J. Clin. Pathol. 66*, 753 (1976).

128. Koenig, R. J., Peterson, L. M., Jones, R. L., Sandek, C., Lehrman, M. and Cerami, A. : *N. Engl. J. Med. 295*, 417 (1976).

129. Trivelli, L. A., Ranney, H. M. and Lai, H. T. : *N. Engl. J. Med. 284*, 353 (1971).

130. Vesterberg, O. : in Radola, B. J. and Graesslin, D. (eds.), : *Electrofocusing and Isotachophoresis*, de Gruyter, Berlin, p. 148 (1977).

131. Delmotte, P. : *Z. Klin. Biochem. 9*, 334 (1971).

132. Stibler, H. and Kjellin, H. G. : *Acta Neurol. Scand. 54*, 119 (1976).

133. Lorincz, L. L. : in Allen, R. C. and Arnaud, P. (eds.), : *Electrophporesis '81*, de Gruyter, Berlin, p. 149 (1981).

134. Confavreu, C., Giannazza, E., Chazot, G., Lasne, Y. and Arnaud, P. : *Electrophoresis 3*, 210 (1982).

135. Robtol, L. : *Clin. Chim. Acta 29*, 101 (1970).

136. Hall, P. W. and Vasiljevic, M. : *J. Lab. Clin. Med. 81*, 897 (1973).

137. Boulton, F. E. and Huntsman, R. G. : *J. Clin. Pathol. 24*, 816 (1971).

138. Hultberg, B., Ockerman, P. A. and Norden, N. E. : *Clin. Chim. Acta 52*, 239 (1974).

139. Kellar, K. L., Vogler, W. R. and Kinkade, T. N. : *Proc. Soc. Exp. Biol. Med. 150*, 766 (1975).

140. Beeley, J. A. : *Arch. Oral. Biol. 14*, 559 (1969).

141. Bennick, A. and Cornell, G. E. : *Biochem. J. 123*, 455 (1971).

142. Chisholm, D. M. Beeley, J. A. and Mason, D. K. : *Oral Surg. 35*, 620 (1973).

143. Pronk, J. C. : in Radola, B. J. and Graesslin, D. (eds.), *Electrofocusing and Isotachophoresis*, de Gruyter, Berlin, p. 359 (1977).

144. Bustos, S. E. and Fung, L. : in Allen, R. C. and Arnaud, P. (eds.), *Electrophoresis '81*, de Gruyter, Berlin, p. 317 (1981).

145. Maurer, H. R. : *Disc Electrophoresis*, de Gruyter, Berlin, p. 147 (1971).

146. Wadstrom, T. and Smith, C. J. : in Arbuthnott, J. P. and Beeley, J. A. (eds.), *Isoelectric Focusing*, Butterworths, London, p. 152 (1975).

147. Latner, A. L., Parsons, M. E. and Skillen, A. W. : *Biochem. J. 118*, 299 (1970).

148. Alpert, E., Coston, R. L. and Drysdale, J. W. : *Nature* (London), *242*, 194 (1973).

149. Burnett, J. B. and Seiler, H. : *J. Invest. Dermatol. 52*, 199 (1969).

150. Yam, L. T., Li, C. Y. and Lam, K. W. : *N. Engl. J. Med. 284*, 357 (1971).

151. Hennis, H. L., Allen, R. C., Hennigar, G. R. and Simmons, M. A. : *Electrophoresis 2*, 187 (1981).

152. Allen, R. C., Gale, G. R., Oulla, P. M. and Gale, A. O. : *Bioinorg. Chem. 8*, 83 (1978).

153. Allen, R. C., Gale, G. R. and Simmons, M. A. : *Electrophoresis 3*, 114 (1982).

154. Saravis, C. A., Cunningham, C. G., Marasco, P. V., Cooke, R. B. and Zamcheck, N. : in Radola, B. J. (ed.), *Electrophoresis '79*, de Gruyter, Berlin, p. 117 (1980).

155. Thompson, B. J., Dunn, M. J., Burghes, A. H. M. and Dubowitz, V. : *Electrophoresis 3*, 307 (1982).

156. Leabeck, D. H. and Robinson, H. K. : *FEBS Lett. 40*, 192 (1968).

157. Neumann, H., Moran, E. M., Russell, R. M. and Rosenberg, I. H. : *Science 186*, 151 (1974).

158. Wadstrom, T., Nord, C. E. and Kjellren, M. : *Scand. J. Dent. Res. 84*, 234 (1976).

159. Koster, J. F., Slee, R. G., van der Kei-Van Moorsel, J. N., Reitra, P.J.G.M. and Lucas, C. J. : *Clin. Chim. Acta 68*, 49 (1976)

160. Alhadeff, J. A., Miller, A. L., Wenger, D. A. and O'Brien, J. S. : *Clin. Chim. Acta 57*, (1974).

161. Turner, B. M., Beratis, N. G., Turner, V. S. and Kirschhorn, K. : *Nature* (London), : *257*, 391 (1975).

162. Kahn, A., Hakim, J., Cottreau, D. and Boivin, P. : *Clin. Chim. Acta 59*, 183 1975).

163. Der Kaloustian, V. N., Idriss-Daouk, S. H., Hallal, R. T. and Awdeh, Z. L. : *Biochem. Genet. 12*, 51 (1974).

164. Hayase, K. and Kritchevsky, D. : *Clin. Chim. Acta 46*, 455 (1973).

165. Hultberg, B. : *Lancet 2*, 1195 (1969).

166. Harzar, K. : *Klin Wochenschr. 52*, 145 (1974).

167. Kint, J. A., Dacremont, G., Carton, D., Orye, E. and Hooft, C. : *Science 181*, 352 (1973).

168. Jareki, R., Pogacar, P., Gunther, G. and Klein, H. : *Rechts Med. 67*, 313 (1970).

169. Buck, G. and Grebe, R. : Dissertation - Ruprecht-Karl-Universität, Heildelberg, p. 104 (1976).

170. Young, C. W. and Bitter, E. S. : *Cancer Res. 33*, 2692, (1973).

171. Phillips, G. R. : in Allen, R. C. and Maurer, H. R. (eds.), *Electrophoresis and Isoelectric Focusing in Polyacrylamide Gels*, de Gruyter, Berlin, p. 255 (1974).

172. Thorstensson, A., Sjordin, B. and Karlsson, J. : in Righetti, P. G. (ed.), *Progress in Isoelectric Focusing and Isotachophoresis*, Elsevier North / Holland, Amsterdam, p. 213 (1975).

173. Anido, V., Conn, R. B., Mengoli, H. F. and Anido, G. : *Amer. J. Clin. Pathol. 61*, 599 (1974).

174. Chamoles, N. and Karcher, D. : *Clin. Chim. Acta 30*, 337 (1970).

175. Zail, S. S. and Van Den Hoek, A. K. : *Clin. Chim. Acta 79*, 15 (1977).

176. Allen, R. C., Sannes, P. L., Spicer, S. S. and Hong, C. C. : *J. Histochem. Cytochem. 28*, 947 (1980).

177. Oda, E., Nakashima, K., Shinohara, K. and Miwa, S. : *Clin. Chim. Acta 68*, 93 (1976).

178. Vogel, F. and Altland, K. : in Bora, K. C. *et al.* (eds.), *Progress in Mutation Research 3*, Elsevier Biomedical Press, Amsterdam, p. 143 (1982).

179. Allen, R. C. : in Radola, B. J. (ed.) *Electrophoresis '79*, de Gruyter, Berlin, p. 631 (1980).

180. Kühnl, P. : *Int. Tagg. Ges. for. Blutgruppenkd.*, Hamburg, (1977).

181. Harada, S., Agarwal, D. P. and Goedde, H. W. : *Hum. Gen. 44*, 215 (1978).

182. Turner, B. M., Beratis, N. G., Turner, V. S. and Hirschhorn, K. : *Am. J. Hum. Gen. 27*, 651 (1975).

183. Tariverdian, G., Ritter, H. and Wendt, G. G. : *Humangenetik 11*, 75 (1970).

184. Narayanan, K. R. and Raj, A. S. : in, Radola, B. J. and Graesslin, D. (eds.), *Electrofocusing and Isotachophoresis* 221 (1977).

185. Kömpf, J., Bissbort, S., Gussmann, S. and Ritter, H. : *Humangenetic 27*, 141 (1975).

186. Chen, S-H. and Giblett, E. R. : *Science 173*, 148 (1971).

187. Bark, J. E., Harris, M. J. and Firth, M. : *J. Forensic Sci. Soc. 16*, 115 (1976).

188. Fildes, R. A. and Parr, C. W. : *Nature 200*, 890 (1963).

190. Harris, H., Hopkinson, D. A. and Robson, E. B. : *Ann. Hum. Gen. 37*, 237 (1974).

191. Allen, R. C., Arnaud, P. and Spicer, S. S. : in Allen, R. C. and Arnaud, P., (eds.), *Electrophoresis '81*, de Gruyter, Berlin, p. 167 (1981).

192. Görg, A., Postel, W., Weser, J., Weidinger, S. Patutschnick, W. and Cleve, H. : *Electrophoresis 4*, 153 (1983).

193. Weidinger, S., Schwarzfischer, F. and Cleve, H. : in Stathakos, D. *Electrophoresis '82*, de Gruyter, Berlin, p. 761 (1983)

194. Constans, J., Viau, M. : *Science 198*, 1070 (1977).

195. Latner, A. L., Emee, A. V. : in Righetti, P. G. (ed.), *Progress in Isoelectric Focusing and Isotachophoresis*, Elsevier,North/Holland, Amsterdam, p. 223 (1975).

196. Basset, K., Beuzard, Y., Garel, M. C. and Rosa, J. : *Blood 51*, 971 (1978).

197. Johnson, A. M., Schmid, K. and Alper, C. A. : *J. Clin. Invest. 48*, 2293 (1969).

198. Allen, R. C. and Arnaud, P. : *Electrophoresis 4*, 205 (1983).

199. Utermann, G., Hess, M. and Vogelberg, K. H. : in, Radola, B. J. and Graesslin, D. (eds.), *Electrofocusing and Isotachophoresis* de Gruyter, Berlin, p. 281 (1977).

200. Alper, C. A. and Johnson, A. M. : *Vox Sang. 17*, 445 (1969).

201. Alper, C. A. : *J. Exper. Med. 144*, 1111 (1976).

202. Cooke, R. J. : *Electrophoresis 5*, In press (1984).

203. Righetti, P. G. and Bianchi Bosisio, A. : *Electrophoresis 2*, 65 (1981).

204. Stegemann, H. and Loeschke, C. : *Index of European Potato Varieties*, Mitt. Biol. Bundesanstalt, Berlin, Heft 168, 215 pp. (1976).

205. Drawert, F. and Görg, A. : *Z. Lebensm. Unters. Forsch. 159*, 23 (1975).

206. Rosengren, A., Bjellqvist, B. and Gasparic, V. : in Radola, B. J. and Graesslin, D., (eds.), *Electrofocusing and Isotachophoresis*, de Gruyter, Berlin, p. 165 (1977).

207. Righetti, P. G. and Gianazza, E. : in Radola, B. J., (ed.), *Electrophoresis '79* de Gruyter, Berlin, p. 23 (1980).

208. Righetti, P. G., Krishnamoorthy, R., Gianazza, E. and Labie, D. : *J. Chromatogr. 166*, 455 (1978).

209. Righetti, P. G. and Gianazza, E. : *Prot. Biol. Fluids 27*, 711 (1979).

211. Righetti, P. G., Gacon, G., Gianazza, E., Lostanlen, D. and Kaplan, J. C. : *Biochem. Biophys. Res. Commun. 85*, 1575 (1978).

212. Constans, J., Viau, M., Gouaillard, C., Bouisson, C. and Clerc, A. : in Radola, B. J. (ed.), *Electrophoresis '79* de Gruyter, Berlin, 701 (1980).

213. Laurent, P., Potempka, L. A., Gewurz, H, Fiedel, B. A. and Allen, R. C. : *Electrophoresis 4*, 316 (1983).

214. Valenti, L., Gianazza, E. and Righetti, P. G. : *J. Biochem. Biophys. Methods 3*, 323 (1980).

215. Horejsi, V. : *J. Chromatogr. 178*, 1 (1979).

216. Allen, R. C. and Arnaud, P. : in Allen, R. C., Bienvenue, J., Laurent, P. and Suskind, R. M., (eds.), *Marker Proteins in Inflammation* de Gruyter, Berlin, p. 185 (1982).

217. Fawcett, J. S. : in Arbuthnott, J. P. and Beeley, J. A. (eds.), *Isoelectric Focusing*, Butterworth, London, p. 23 (1975).

218. Fawcett, J. S. : in Catsimpoolas, N., (ed.), Isoelectric focusing, Academic press, New York, p. 173, (1976).

219. Radola, B. J. : in Catsimpoolas, N. (ed.), *Isoelectric Focusing*, New York, p. 119, (1976).

220. Chrambach, A. and Nguyen, N. Y. : in Righetti, P. G., Van Oss, C. J. and Vanderhoff, J. W. (eds.), *Electrokinetic Separation Methods*, Elsevier/North Holland, Amsterdam, p. 337, (1979).

221. Righetti, P. G., Giananaza, E. and Ek, K. : J. Chromatogr. 184, 415, (1980).

222. Jonsson, J. and Rilbe, H. : *Electrophoresis 1*, 3, (1980).

223. Bier, M., Egen, N. B., Allgyer, T. T., Twitty, G. E. and Mosher, R. A. : in Gross, E. and Meinenhofer, J. (eds.), *Peptides: Structure and Biological Function*, Pierce Chemical, Rockford, IL, p. 79 (1979).

224. Frey, M. D. and Radola, B. J. : *Electrophoresis 3*, 216 (1982).

225. Görg, A., Postel, W. and Westermier, R. : *Anal. Biochem. 89*, 60 (1978).

226. Radola, B. J. : in Radola, B. J. (ed.), Electrophoresis '79 de Gruyter, Berlin, p. 79 (1980).

227. Kinzkofer, A. and Radola, B. J. : *Electrophoresis 4*, 408 (1983).

228. Radola, B. J. : in Radola, B. J. (ed.), *Elektrophorese '80* Technische Universität, München, p. 43 (1980).

229. Cook, R. and Saravis, C. A. : Marine Colloids Instruction Leaflet (1979).

230. Jeppsson, J. O., Franzen, B. and Gaal, A. B. : in Radola, B. J. (ed.), *Electrophoresis '79* de Gruyter, Berlin, p. 655 (1980).

231. Altland, K. and Kaempfer, M. : *Electrophoresis 1*, 57 (1980).

4. MULTIPARAMETER TECHNIQUES

4.1. Two-Dimensional Electrophoresis

Two-dimensional electrophoresis has become an ever wideningly used tool in biology and medicine for the analytical separation of macromolecules. Since the first reports on isoelectric focusing under dissociating conditions in the first dimension and and denaturing SDS-Page in the second dimension, by O'Farrell (1), Klose (2) and Scheele (3) in 1975, there has been a literal explosion of publications. Already over 3000 publications have appeared in the literature. As in any rapidly developing applied field, there is a considerable controversy concerning the ideal, or even optimal marriage of theory and practice of charge fractionation methods in the first dimension and size separation in the second dimension. While such controversies are healthy and bode well for the continued improvement of the technique, at present it is confusing, if not bewildering, for the neophyte as well as the more experienced investigator to attempt to extract the wheat from the chaff.

We have attempted to maintain objectivity in presenting this area and comments, critical or otherwise, on the many facets of the procedure are solely aimed at bringing forth those arguments, both practical and theoretical, pertinent to the overall performance of the technique. The rationale for the study of native protein versus denatured subunits is, of course, dependent on the study at hand. We shall attempt to trace the development of the technique, present concepts and examine the methodological steps involved in two- dimensional as well as in more complex multiparameter procedures.

4.1.1. Development

Since the original two-dimensional electrophoretic separations, based on the physical-chemical parameters of charge and size, by Smithies and Poulik in 1956 (4), the term has come to be inclusive for a number of succeedingly and increasingly complex techniques. Systems utilizing more than one physical chemical parameter *i.e.*, charge and size of a macromolecule separated solely by electrophoretic methods may be considered, in the classical sense of the term, a two-dimensional technique.Such charge- size isomerisms have been utilized in paper-starch gel, cellulose acetate -starch gel, cellulose acetate-PAGE and PAGIF, agar-SDS-PAGE and PAGIF-SDS-PAGE.

The practical realization of the two-dimensional technique as a high resolution tool in biological and biomedical research required a series of key developments to achieve its present level of sophistication. In 1969, Margolis and Kendrick (5) combined Disc electrophoresis with pore gradient acrylamide gel electrophoresis, while Dale and Latner (6) combined isoeletric focusing and polyacrylamide gel electrophoresis for the two-dimensional separation of serum proteins andMako and Stegemann (7) used a similar system for mapping potato proteins. Fortuitously in the same year, Weber and Osborn (8) published a critical study for the further development of the two-dimensional technique in their report on the reliability of molecular weight determination by sodium dodecyl sulfate-polyacrylamide electrophoresis. Thus, utilization of denatured proteins and subunit polypeptide could now be resolved accurately on the basis of molecular weight and a suitable second dimension was now available.

The two-dimensional system of Kaltschmidt and Wittman (9) on polyacrylamide gel for the finger printing of ribosomal proteins led Mets and Bogorad (10) improving upon the resolution of such ribosomal maps by utilizing urea in the first dimension combined with SDS-PAGE in the second dimension. This work was the first indication that SDS would rapidly complex to proteins in the presense of urea without heating. Although, whether adequate SDS saturation under such conditions in the absense of heat is complete or not, still requires appropriate experimental verification. At this time, other investigators had or were utilizing PAGIF in the presense of urea in a first dimension and SDS-PAGE in the second dimension, Barret and Gould with chromosomal proteins (11), Stegemann *et al.* (12) with potato proteins and Bahakdi *et al.* (13) with erythrocyte membranes.

As Klose (14) has pointed out: "in spite of the fact that all important technical advances were known and all investigators had emphasized the high resolving power of the *new* two-dimensional electrophoresis, no attempt was made to demonstrate the potential resolution of this technique by separating a highly complex protein solution such as all the proteins of a cell or of an organism". It remained for Klose

(2), O'Farrell (1) and Scheele (3) with mouse tissue extracts, *E. coli* and guinea pig pancreas extract respectively, to show that when dissociating isoelectric focusing was followed by denaturing SDS-PAGE some 1100 to 1250 individual polypeptide spots could be demonstrated.

Since 1975, there has been a steady increase of interest in the technique leading to the ISO-DALT system in 1979 by Anderson and Anderson (15, 16), which culminated in the concept of *The Human Protein Index* in 1980 by these two investigators (17). The recent almost explosive development of the technique in biology and medicine finally indicates that electrophoresis has the resolution potential optimistically claimed ,first with the advent of Disc electrophoresis and again with the development of isoelectric focusing. These developments have been the subject of an excellent and extensive review series by Dunn and Burghes (18,19).

4.1.2. Theoretical considerations affecting resolving power

For the optimal use of two-dimensional PAGIF-SDS PAGE it is important to maximize the resolution that can be obtained in both dimensions in order to obtain as much information from a single polypeptide map as possible. To best understand the interacting theoretical and practical parameters affecting optimal resolution, it is best to handle each dimension separately.

4.1.2.1. The first dimension

In most systems such as the ISO-DALT, the isoelectric focusing step has essentially continued to be that as used by O'Farrell (1) and Klose (2) where the separation is carried out in gel rods, rather than in gel slabs as was done by Scheele (3). Normally pH gradients of 3 to 10 are used in gel rods of 16 cm or more in length in the presense of 8-9.5 M urea. In such a system there is considerable pH gradient instability associated with the cathodal region (20,21) and the pH gradients do not extend significantly above pH 7.0. This has resulted in the development of non-equilibrium systems for the first dimension or non-equilibrium pH gradient focusing (NEPHGE) or transient state focusing (22). However, there are at least two drawbacks to such systems: 1), that serum proteins, for reasons that are yet unclear, do not perform well in non-equilibrium systems (21); 2), the position of a spot under non-equilibrium conditions on polyacrylamide gels is a function of both charge and size. The latter would make assignment of pI value more difficult for those interested in accurately cataloging proteins using pI as one of the parameters, although, the use of internal carbamylation chain standards can overcome this difficulty.

Tracy (20) has approached this problem for the separation of basic myeloma proteins by a modification of ampholytes and the inclusion of arginine and lysine in the system. However, voltage drop studies indicated that field strength inhomogeneity was still present, although; the high resolution region was extended from about 7.0 to a pH of about 9.5. Burghes *et al.* (23) have approached the problem from a different direction by utilizing the greater resolution potential of thin flat slab isoelectric focusing at high voltage gradients. The rationale is similar to that reported earlier by Allen (24) and Allen and Arnaud (25) to achieve steady state focusing at both the anodal and cathodal ends of the gel by increasing the voltage gradient. A comparison of the rod gel PAGIF and horizontal thin layer separation on the overall resolution of the 2-D pattern is shown in Fig. 1a and 1b from the work of Burghes *et al.* (26). It is evident that resolution is markedly enhanced in both the cathodal and anodal end of the 2-D map from the thin-layer PAGIF gels where the higher voltage gradients are employed on first dimension slabs (see Chapter 3, section 3.4).

Based on previous reports that various commerical ampholytes give better resolution in different pH regions of focusing gels (26-29) they also used ampholyte tailoring of mixtures of commercially available ampholytes to improve overall resolution further. This also resulted in their system giving straight pH lines which are important for obtaining maximal resolution or minimal spot width in the second dimension.This study supports the earlier work of Scheele (3) and Görg, *et al.* (30) on the advantages of horizontal slab electrophoresis to produce improved resolution in the second dimension not achievable with cylindrical rods. In this respect, it is interesting to note that in some 80 papers of PAGIF-SDS PAGE with 330 full 2-D maps and 400 partial maps, that only in the original papers by O'Farrell and Scheele (1,3) were duplicate gel rods or slabs of the PAGIF dimension shown. One separate rod with slanted bands was shown by itself (29).

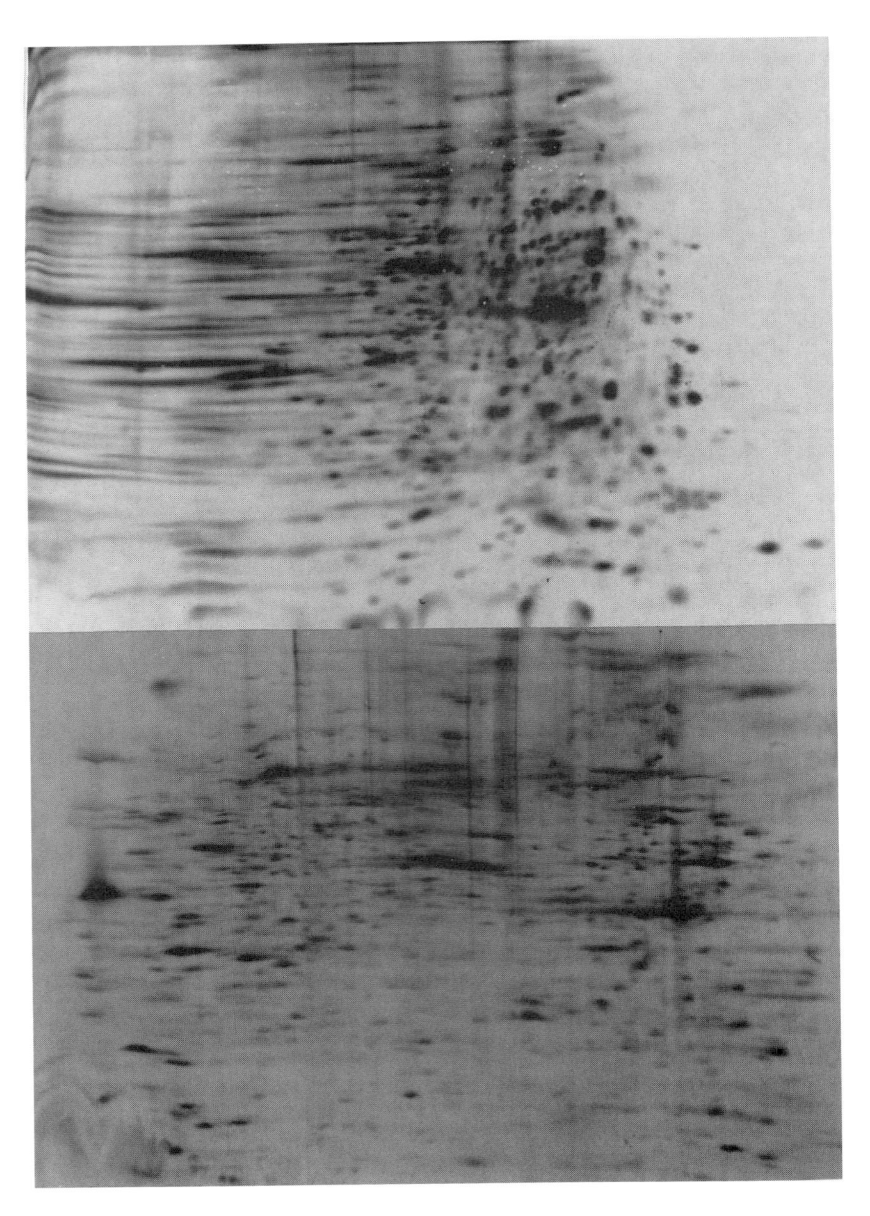

Fig. 1. Two-dimensional separation of skin fibroblasts with first dimension separation on a gel rod A and first dimension on a thin-layer slab B. (from Burghes *et al.*) (24).

Furthermore, only 5 pH scales were indicated in this literature where gel rods were used. Practitioners of the technique utilizing gel rod PAGIF have relied in the main on charge isomer standards and have apparently tacetly accepted that the first dimension produced absolutely parallel discs and need not be further optimized.

The controversy of gel rods versus horizontal slabs in the first dimension is probably not ended, however, the more recent reports by both Burghes, *et al.* (30) and Görg, *et al.* (31) and the earlier work of Scheele (3) offer a potentially important improvement for optimizing resolution in the first dimension based on the theory of isoelectric focusing. Nevertheless, the findings of Kinzkofer and Radola (32) (see enzyme visualization Chapter 5) that the diffusion that occurs in the first several minutes following focusing is sufficient to seriously interfere with resolution of low molecular weight macromolecules, suggests that none of the present handling methods between the first and second dimension approaches the speed that actually is needed.

Unfortunately there is a paucity of information in the literature on resolution potential or reproducability of PAGIF in polyacrylamide gel rods. O'Farrell indicated a variation of 1.2 per cent in the rod gel PAGIF (1) which translates into a possible shift of 0.4mm in a spot. With band thicknesses of 0.1mm possible separated by as little as 80μ from one another (22). The following question must be raised: is this a sufficient degree of resolution in the first dimension to study adequately biologic, physiologic, pathologic and genetic changes? Tracy *et al.* (33) find present resolution adequate to allow development of methods for quantitative studies of serum from 2-D gels, but present clinically accepted limits may be too imprecise to fully exploit the potential of this technique. This has been addressed further by Young and Tracy (34) in a recent overview on the usefulness of two-dimensional methods in clinical studies.

4.1.2.2. The Second Dimension SDS-PAGE

This step in the procedure may be subdivided into the system using homogeneous gel and those employing gradient gels. The latter provides the ability to engineer the gradient and to take fuller advantage of the MZE buffer systems stacking and unstacking effects on resolution.

Experimental concave gradients have dominated in this field in conjunction with the Glycine-Chloride MZE buffer system (1,14,15,20,30,31) and indeed the Burghes, *et al.* (30) modified exponential gradient is actually tailored to optimize resolution of fibroblast proteins.

Two salient points that appear in the literature directly apply to SDS-PAGE, apparently overlooked in most of the 2-D literature, to further optimize resolution in the second dimension. These are given more detail in the PAGE-SDS section but are again emphasized here. The first of these is the observation by Wykoff, *et al.* (35), that SDS migrates with a mobility higher than SDS-protein complexes in a restrictive gel. It will then necessarily overtake the zones of protein in the resolving gel if present in the sample and upper buffer. It is thus, unnecessary to add SDS to the stacking or separating gel. Omission of SDS has the further advantage of ensuring that conditions of polymerization elaborated for the non-SDS case remain applicable to SDS-PAGE and that stack broadening with resultant loss of resolution is minimized. The second is that since SDS-proteins are not titratable between pH 7 and 10, mobility is not variable within this range and stacking is achieved by molecular sieving only. Therefore, MZE of SDS-proteins does not require an operative pH between stacking and resolving gel and can utilize either the zeta or pI phase of any MZE buffer system in the neutral to alkaline pH range. This provides a further simplification of the system and permits additional MZE buffer systems which may enhance resolution of complex mixtures to be employed (see molecular weight determination Chapter 2, section 2.3.2.).

In many cases the investigator may wish to utilize non-denaturing conditions in the second dimension for the study of native proteins enzymes or direct immuno-printing. Rothe and Parkhandaba (36,37) have systematically investigated and reviewed both homogeneous and gradient systems for this purpose, as well as listing two-step methods used in the determination of molecular weights of native proteins in gradient gel electrophoresis (see Chapter 2, section 2.3.1.).

Görg, *et al.* (31) have also utilized non-denaturing conditions in the second dimension with horizontal electrophoresis on thin (0.5mm) exponential, concave gradient gels. This allows direct enzyme determination and should also provide for direct immunoprint techniques. Unfortunately, these authors do not indicate whether the replicate print technique of Narayanan and Raj (38) and Pretch, *et al.* (39) could be used to expand further the information potential of a single 2-D map (see enzyme visualization, chapter 5 section 5.6.1.1.).

4.1.2.3. Sample treatment

Handling of the sample for the PAGIF step ranges from simple buffer extraction for native proteins to a variety of empirically arrived at buffer-detergent-urea mixes for denaturation. Representative sample treatment mixes for a variety of biological samples are shown in Table 1 on page 158.

4.1.2.4. Solubilization mixes

The advantages and disadvantages of using an anionic detergent as a denaturant have been investigated in gel rods by Klose and Feller (40). They found that, under all conditions tested, SDS in the protein sample had detrimental effects in the protein patterns, the most serious of which was a loss of protein during PAGIF. These authors, as well as Horst, *et al.* (41), have observed some proteins which were usually lost when no SDS was present in the samples. In the case of samples applied at the anodic end, the bulk of proteins migrated, apparently in association with SDS in the opposite direction and were lost. With application of the sample at the cathodal end of the rod, when the samples were placed in Sephadex 200 first to facilitate gel entry, a greater portion of proteins entered the gel. However, background staining occured in the gel rod leading to streaks in the second dimension SDS-PAGE separation. Klose and Feller (40) have concluded that the SDS presumably did not dissociate completely for all proteins and prevented them from entering the gel. This was apparently a disadvantage of loading at the cathodal end, as shown in Fig. 2a and 2b). Also in these experiments NP-40 did not promote displacement of SDS from the proteins during PAGIF in the presence of urea (the use of CHAPS instead of Triton X 100 may overcome this problem). Heating of crude protein samples containing SDS was also found to increase the loss of proteins. Negative effects of sample loading were also described by Wilson, *et al.* (42) and Stegemann (43).

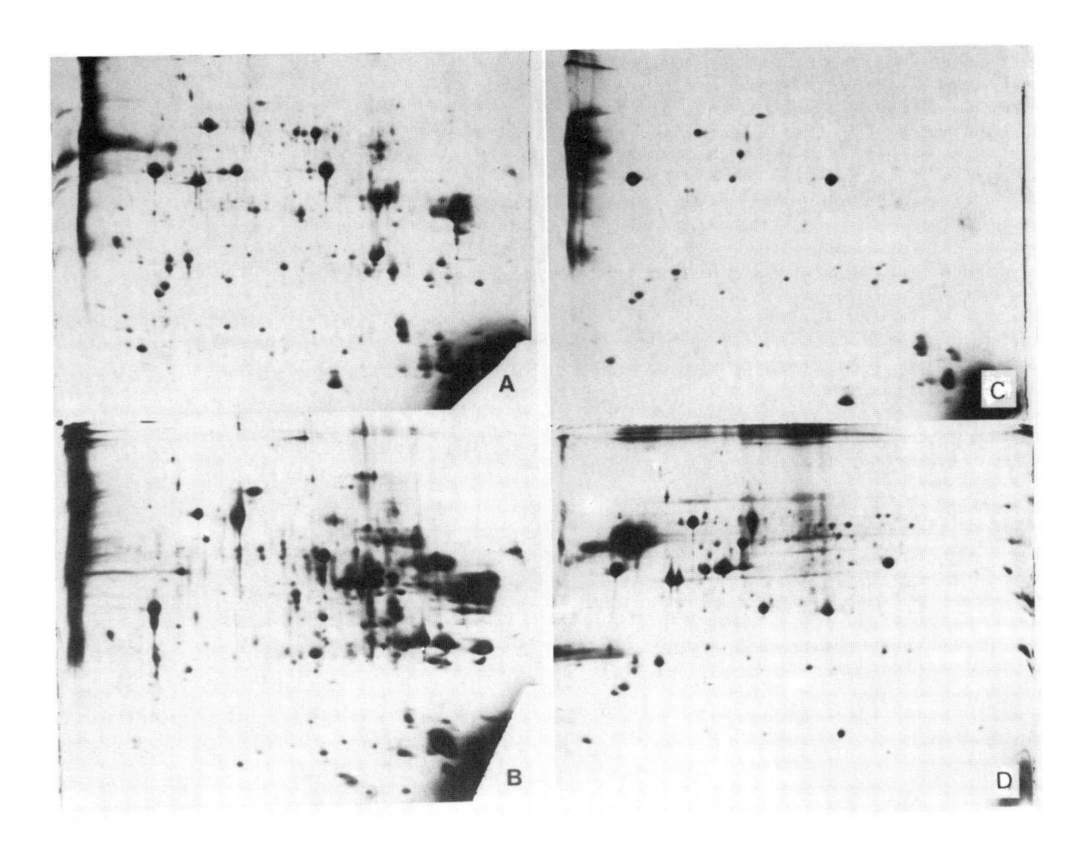

Fig. 2. Cytosol proteins obtained from the organs of inbred mice. (A) Proteins of brain cytosol, (B) proteins of liver cytosol stained with Coomassie Brilliant Blue R250. The same samples are shown in (C) and (D), but with the samples containing 0.2 % SDS. All samples in the first dimension were applied to the anode side, left, and migrated to the cathode, right. (from Klose and Feller) (40)

On the other hand Burghes, *et al.* (23) using skin fibroblasts from both normal and Duchennes Muscular Distrophy patients, in a comparison of solubilization procedures (given on page 158) indicated that the addition of SDS followed by competition with non-ionic detergent (44,45) produced patterns similar to those used employing a Urea/NP-40 mixture, except that there was less material at the point of application with both anodal and cathodal loading and that there was an increase in 2 bands in the middle of the gel.

The results of the two PAGIF procedures, presented above, suggest that loading the SDS denatured samples directly in the surface of partially prefocused horizontal slab gels obviates the problems encountered with protein loss of entry in gel rod systems. The considerably higher voltage gradients used on the former, not only increases resolution, but also may facilitate SDS protein entry into the gel.

This report suggests further that the concept of "Competition" is misleading and that formation of a mixed micelle (46,47) is far more likely to occur. No matter how much NP-40 is added, every micelle will contain some SDS and thus, the negatively charged micelles will migrate to the anode and be removed from the gel. The previously solubilized proteins will then be exposed to a detergent-poor environment and those which depend on the presence of detergent for their solubility will tend to precipitate. This effect has been illustrated by Murkegee (48). Mixed micelles of non-ionic and Zwitterionic detergents have been reported to have increased solubilizing ability (49-51). Burghes, *et al.* (23) found that a Sulfobetaine/NP-40 mixture with (Buffer 4, Table 1) or without (Buffer 5, Table 1) the addition of urea was less effective than the urea/ NP-40 combination and that high levels of urea caused percipitation of Zwitterionic detergents.

Such studies on the various choices of detergents, anionic, nonionic or Zwitterionic suggest, in the presence of dissociating urea systems, that many of the observed difficulties may be due to the addition of excess competing components into the sample mix. Additionally, a comparison of gel rods with horizontal thin gel slabs provides an environment free of direct contact with catholyte and anolyte, which may interact adversely with mixed ionic-nonionic systems.

SOLUBILIZATION MIXES

	MATERIAL	SDS	TEMP.	TIME MIN.	OTHER	UREA	DETERTENT	REDUCING AGENT	AMPHOLYTE	REF.
	E. Coli					9.5M	2%	5%, βME	1.6% pH 5-7	(1)
	Cytosol Liver					9M		5%, βME	2% pH 5-7	(40)
	Brain					9M		5%, βME	2% pH 5-7	
	Membrane Liver					9M		5%, βME	2% pH 5-7	
	Membrane Brain					9M		5%, βME	2% pH 5-7	
(1)	Serum-plasma	2%	95°C	5 min	10% Glycerol		CHES 0.05m	1%, DTT		(56)
(2)	Solid tissue Tissue Culture Cells, Lymphocytes				1mM PMSF	9M	4% NP-40	5%, βME	2%	
(3)	Urine					9M		5%, βME	2% pH 3-10	
(4)	Muscle					9M	4% NP-40	1%, DTE	2% pH 3-10	
(1)	Skin Fibroblasts				0.1M or 0.1M Arginine	9.2M	3% NP-40	5%, βME	4% pH 3-10	(30)
(2)	"				5Mm - Na_2CO_3	9.2M	3% NP-40	5%, βME	4% pH 3-10	
(3)	"	2%			0.1mM Arginine			10%, βME		
(4)	"				0.03M Lysine		4% NP-40 0.16M, SB14	2M EDTA 5%, βME	6.8% pH 3-10	
(5)	"	4 above plus				4M Urea	5.3% NP-40	5%, βME	6.8% pH 3-10	
		3 above plus				6M	0.19M, SB14			
	Lymphocytes	2%	95°C	5 min	20% Glycerol	9M	2% NP-40	5%, βME	2% pH 3-10	(52)
	Saliva	2%				9M	2% NP-40		3.6% PH 3.5-10	(53)
	Hair Roots					9M		5%, βME	1% pH 5-7	(54)
	Intestinal	Al above solid tissues								(55)

4.1.2.5. Equilibration, application and protein transfer

Following isoelectric focusing gels are normally equilibrated in pH 6.8 Tris-chloride buffer containing 5 per cent β mercaptoethanol (BME) or 1.25 per cent dithioerythritol (DIE) and 2 per cent SDS (56) or a modified Tris-chloride pH 8.8 20mM dithiothreitol (DTT) 3.0 per cent SDS buffer when stacking gels are not utilized (30). Elimination of the sulfide reducing agents may be desirable since Tracy, *et al.* (57) have reported that this eliminates surface streaks seen on silver stained gels (see Chapter 5., section 5.3.) . Gel rods are normally equilibrated for 30-40 minutes in this solution while thin-layer horizontal gel slabs require only 5-10 minutes and ultrathin-layer gels only 2 minutes. An obvious advantage here in the thin slab gels is that there is far less opportunity for diffusion to take place in the gel prior to application to the second dimension.

Jackle, *et al.* (58) and Görg *et al.* (31) have fixed the first dimension in 50 per cent methanol, 10 per cent acetic acid and then stained the ultrathin-layer PAGIF gel with Coomassie Brilliant Blue G250. This procedure effectively immobilizes the proteins and allows observation of their entry into the second dimension gel. There is no apparent effect on the SDS-PAGE separation. However, this is not true for Coomassie Brilliant Blue R250. On restaining the PAGIF gel slab after SDS-PAGE, no proteins detectable with Coomassie Brilliant Blue R250 remained. On similar restaining of gel rods there remained diffuse staining in the center, which these authors suggest causes vertical streaking. Whether this is due to proteins not included in the initial leading and trailing ion stack when no stacking gel is utilized with gel rod system, or some other reason has not been investigated. On the other hand, the acid alcohol fixation of ultrathin-layer gels has been shown by Radola (59) to leach proteins during fixation. Further definitive studies are required to verify completely the efficacy of this novel approach.

Another promising new development to solve the problem of diffusion in the first dimension, occuring during the tranfer step, is the use of Immobiline. Here the ampholytes and proteins at their isoelectric point should remain reaonably stable during the tranfer step. Additional advantages claimed are the following: 1), the cathodic drift is completely abolished; 2), they give higher resolution and higher loading capacity and 3), they present a milieau of known and controlled ionic strength (60). Their use in controlled studies of 2-D electrophoresis is as yet insufficient to judge the value of this material.

4.1.2.6. Contact of first and second dimension gels

Good contact of the PAGIF and SDS-PAGE gel over the entire length of the rod-slab or slab-slab sandwich is required to achieve optimal electrical conductance. Poor SDS PAGE will result from an uneven boundary in the initial stacking phase, if gaps in the gel-gel contact are present. This is accomplished more easily using gel rods placed on vertical slabs, by simply sealing the two together with appropriately buffered agarose. Flat slab PAGIF to vertical SDS-PAGE gel sealing has been recently accomplished by the use of low melting point agarose from Marine Colloids (FMC Corporation - see Chapter 1) followed by sealing with normal 1.0 per cent agarose in cassettes designed with an offset plate (30). Ultrathin-layer slab PAGIF dimension gels have been layed in slots formed in horizontal thin-layer SDS PAGE gels (31) with apparent good success. Chait (61) has shown that an additional top pressure plate is required for optimal resolution in the second dimension in such systems.

In the horizontal PAGIF - Horizontal SDS-PAGE System of Görg *et al.* (31), have reported also that successful gel to gel contact may be accomplished by casting a slot into the second dimension 0.5mm thick gradient SDS gel and then inserting the first 125µ thick slab into the slot. The ratio of gel thicknesses between the first and second dimension gels here may play a role as reported by Altland in his 2-1D system (62),who found a ratio of 1:3 can not be exceeded. Other techniques such as the agarose IEF in the first dimension and SDS-PAGE in flat slab horizontal system, appear to work reasonably well for some purposes, but do not give results comparable to the acrylamide gel methods discussed above.

4.1.2.7. Fixation and Staining

Fixation and staining are interdependent processes upon which the sensitivity of the method relies. It is best to fix gels in 20 per cent TCA as recommended by Kinzkofer and Radola (32), Merril, *et al.* (63)

and Allen (24) as discussed at length in Chapter 5, section 5.3.2.1.

Staining procedures that have been employed for denaturing 2-D electrophoresis are as follows: Coomassie Brilliant Blue R250, silver stains and autoradiography, comprise the most frequently used stains. These are discussed further under component localization also given in Chapter 5.

4.1.2.8. Reproducibility of 2-D electrophoresis

While many will agree that the optimization of the two dimensions taken independently has not been reached, there is remarkable agreement between gels within and even between laboratories. Reproducibility, determined by densitometric analysis, is less than a spot width (0.5 - 2.0 mm) Taylor *et al.* (64) as illustrated in Fig. 3. Visual illustration of reproducibility, utilizing genetically controlled polymorphisms (65), may be made with a good degree of certainty also as shown in Fig. 4. Such reproducibilty provides an excellent opportunity to study and detect genetic polymorphisms, particularly between father-mother- child, as reported by Skolnick *et al.* (66). On the other hand, irregardless of reproducibility, the wide range gradients normally used make phenotyping and subtyping of phenotypes considerably more difficult as illustrated in Fig. 5, which compares serum 2-D resolution of alpha$_1$-antitrypsin with that obtained by isoelectric focusing on a narrow range pH 3.5 to 5.0 gradient. While single dimension charge based separation methods are quicker and better resolved, they will not detect mutations or genetic polymorphisms based on hydrophobic characteristics of proteins, which may account for two thirds of such polymorphisms. Therefore, the reproducible resolution of 2-D offers the best opportunity to detect polymorphisms, while PAGIF presently offers the better method of phenotyping those polymorphic proteins where charge expression is available for the separation process.

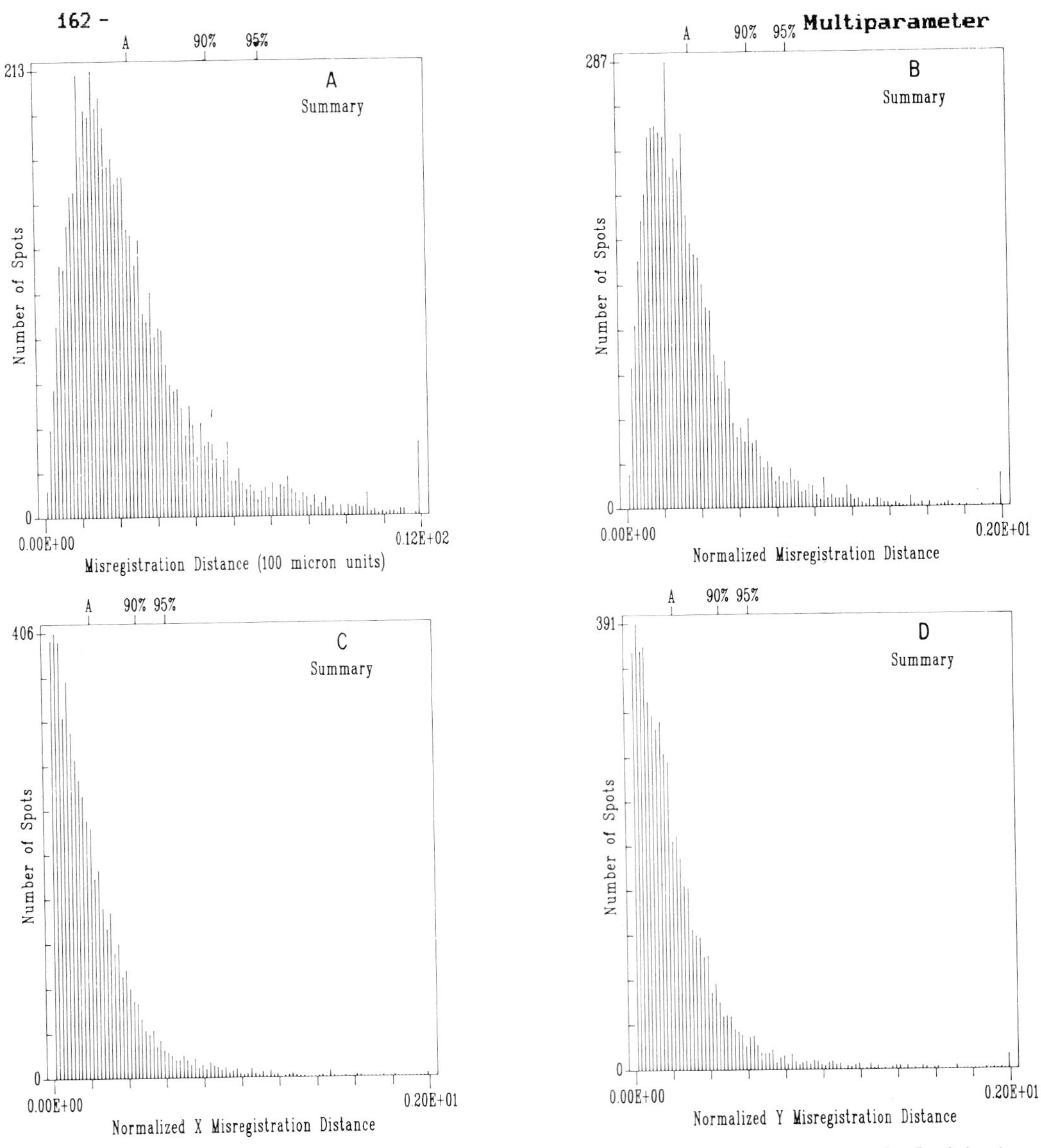

Fig. 3. Histograms for assessing the registration of an entire set of 15 object patterns with a master pattern. (A), distribution of misregistration distance in 100 μmunitsB), normalized misregistration distance; (C),normalized along x-axis; (D), nomalized along y-axis. (from Taylor *et al.*) (64).

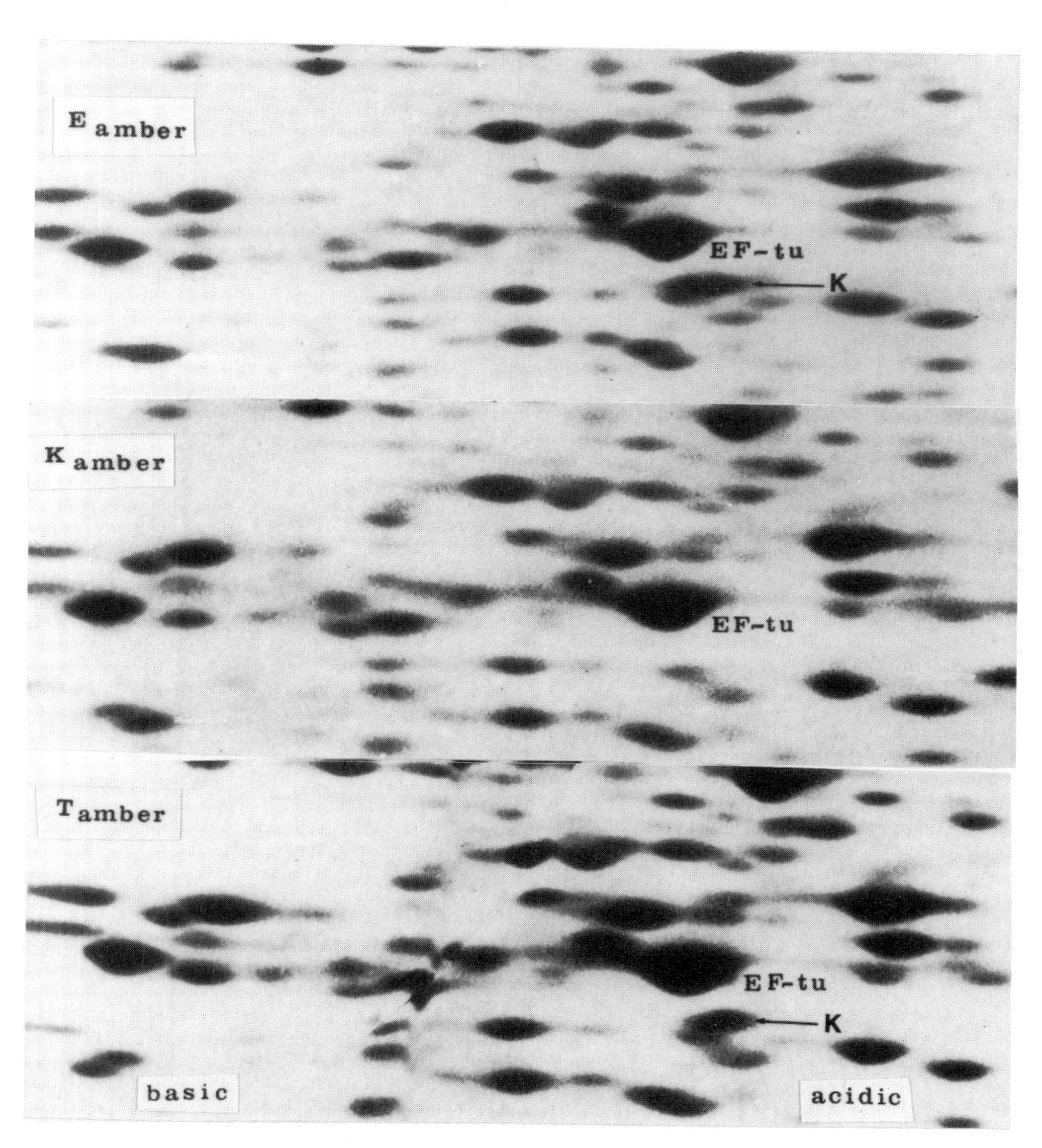

Fig. 4. Spot location reproducibility demonstrated with genetic mutants. (courtesy of Dr. Carl Merril)

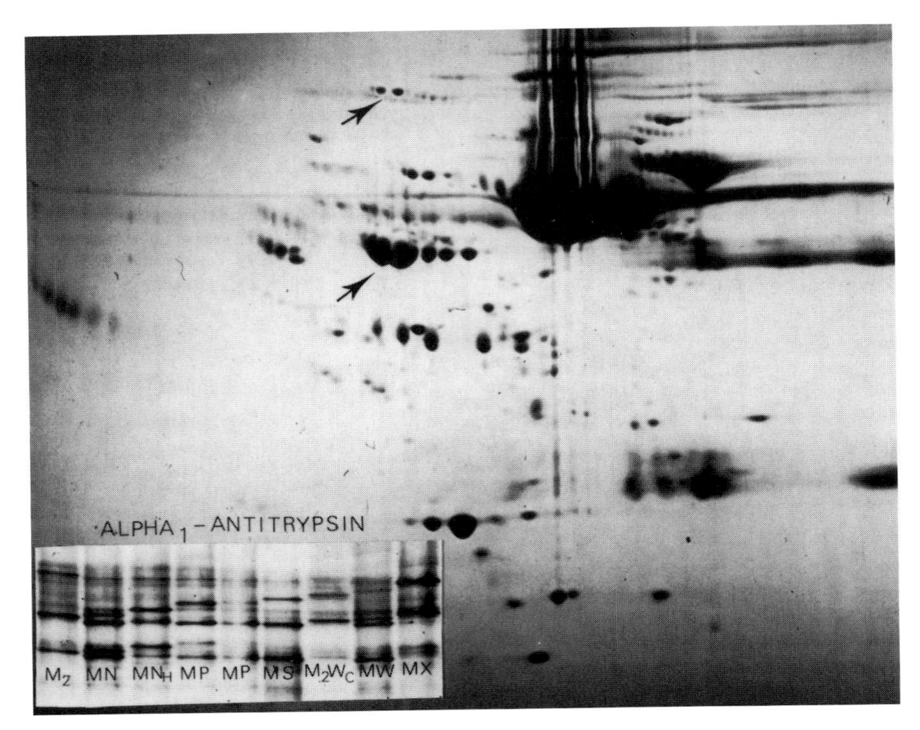

Fig. 5. Resolution of alpha$_1$-antitrypsin comparison between 2-D and PAGIF on a 3.5 -5 pH gradient. The antitrypsin is marked by the arrow on the 2-D gel. (2-D separation courtesy of Dr. N. L. Anderson) (from Allen) (66)

4.1.2.9. Automated analysis

More precise analysis of 2-D gel patterns may be made by computer adjusted maps. A number of such programs have been written for the automated scanners presently in use. Pattern stretching described by Taylor *et al.* (67), R-spot analysis described by Lemkin and Lipkin (68), semi-automatic matching by Miller *et al.* (69) and vector graphing with Gabriel graphs for neighborhood groups, in studying genetic mutation, described by Skolnick (70) are four elegant examples of the automated programs developed for this purpose.

The resolution presently available has allowed a number of disease conditions to be studied with this procedure, as shown in Table 2.

Table 2. Disease conditions studied by 2-D electrophoresis

Disease	Sample	1st-2nd	Method	Ref.
Cystic fibrosis	Saliva	D/D	O'Farrell	(53)
Duchenne muscular dystrophy	Fibroblasts	D/D	ISO-DALT Slab PAGIF DALT	(30)
Fetal Malformation	Amniotic fluid	N/D	Slab PAGIF DALT	(29)
Lesh-Nyhan syndrome	Lymphocytes erythrocytes	D/den	O'Farrell	(64)
Down's syndrome	Fibroblasts	D/D	O'Farrell	(71)
Muliple sclerosis Huntington's Alzheimer's Joseph's	Sclerotic plaques Gliosed caudula Putamen Brain	N/D	NEPHGE Comings	(72-75)
Myeloma proteins	Serum Serum	D/D N/D	ISO-DALT Mod. ISO-DALT	(76) (20)
Pituitary tumor	Brain	D/D	ISO-DALT	(76)
Intestinal carcinoma	Mucosa	D/D	ISO-DALT Mod. ISO-DALT	(55)
Multiple myeloma	Urine	D/D	ISO-DALT	(77)
Neoplastic sensitive proteins	Normal and transformed amnion cells	D/D	NEPHGE-DALT	(78)
Leukaphoresis	Plasma	N/D	Agarose-CZE-DALT	(79)
Juvenile leukemia	Mononuclear cells	D/D	ISO-DALT	(80)
Infectious mononucleosis	Lymphocytes	D/D	ISO-DALT	(81)
Pancreatic adeno-carcinoma	Panceas extract	D/D	Scheele	(82)

D = Denatured, N = Native protein.

In spite of the breadth of disease studies listed in the overview of clinical applications in Table 2, this technique has, not as yet, contributed the significant advances in diagnostic medicine and genetics expected during its earlier developmental stages. The use of only two separation parameters (charge and size), and a finite gel format on which to resolve all of the proteins and polypeptides present in complex biological mixtures, are the limiting factors. As powerful as this technique has proven to be, there is still a detection limit and a peak capacity limit inherent in all systems as presently employed. Full elucidation of all components in a complex biological sample employing additional parameters, such as colored silver staining and electroblot techniques can extend both detection limits and identification capability.Similarly, isotopic labeling can increase sensitivity and dual labeling biological information. However, there is still the major problem of detecting minor components under 50 µg /dl with 2-D techniques alone.

In the following section techniques employing additional approaches, both practical and investigational are given which initially simplify complex mixtures by biochemical or immunological techniques prior to electrophoretic separation, whether this be heresy in a book on electrophoresis or not.

4.2. Other Multiparameter Techniques

4.2.1. Double one- dimensional electrophoresis

Altland and Hackler (83) have developed a system utilizing immunosubtraction and conventional vertical MZE electrophoresis in a first dimension to fractionate the sample and then applying a strip cut from a specific region of the separation to a second isoelectric focusing dimension in the presence of urea for the study of transferrin and using the same technique Altland *et al.* (84) have demonstated the usefulness of this first simplification-step technique in a number of genetic polymorphisms and clinical applications.

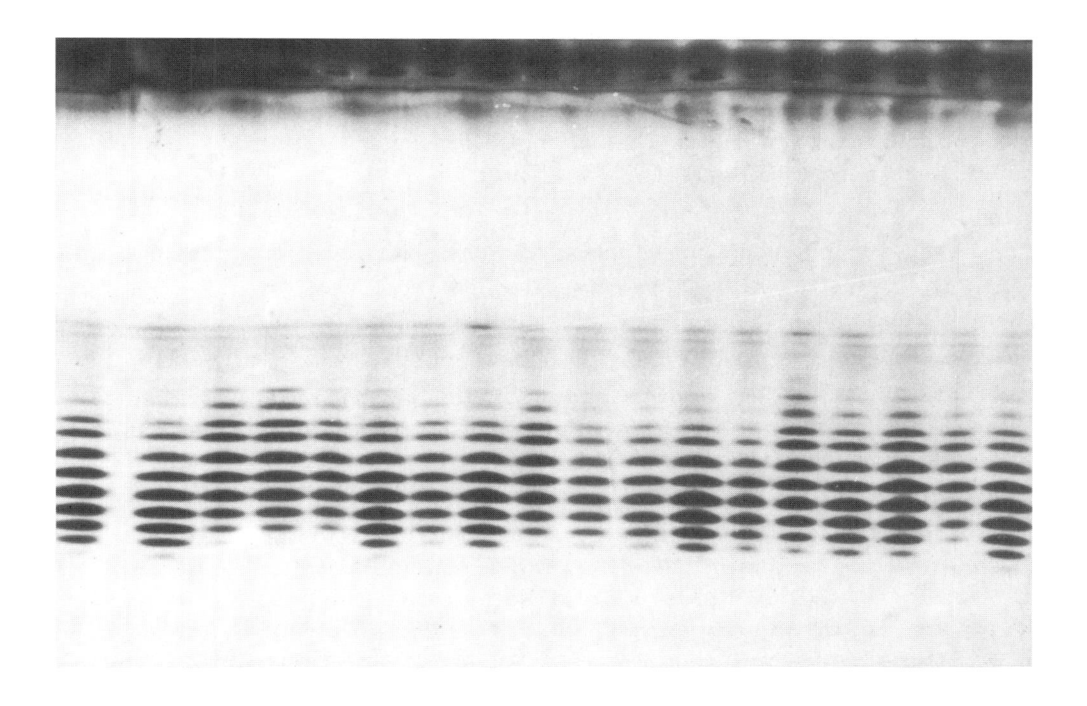

Fig. 6. Separation of acid $\alpha_1$glycoprotein showing normal and abnormal patients with a shift in the center of density of the allelic products following various types of trauma. Only the upper half of gel run from top to center and bottom to center with 96 samples per gel is shown. The cathodal region shows the application region of the first dimensional strip. (courtesy of Dr. K. Altland)

4.2.2. Immunodeletion followed by 2-D.

Complex mixtures of proteins such as cerebral spinal fluid (CSF), prostatic fluid and amniotic fluid also contain contaminating serum proteins, which can add additional complexity to studies directed to non-serum proteins produced locally that are of diagnostic or biologic significance. Dermer and Edwards (85) have employed commercial antisera to human serum coupled to Sepharose 4B or *Staphylococcus aureus* bacteria as immunoabsorbents to remove serum proteins from body fluids prior to 2-D electrophoresis in order to obtain a simplified and specific pattern. Pearson and Anderson (86) have utilized monoclonal antibody-Protein A-microadsorbent columns to to bind specific antigens from complex protein mixtures. They then eluted the antigens for 2-D analysis by the method of O'Farrell.

4.2.3. Pseudo-ligand affinity chromatography and isoelectric focusing.

Arnaud *et al.* (87) have employed Cibacron Blue Sepharose to remove albumin from native protein 2-D patterns, in order to resolve anti thrombin III. Allen and Arnaud (88) have used combinations of specific antibody and Cibacron Blue Sepharose, Con- A, and Matrix Gel A as serum fractionation procedures prior to separation of the native proteins by PAGIF.

Allen and Arnaud (25) have expanded this approach to multiparameter separations, utilizing pseudo-ligand affinity chromatography on Affi-Gel Blue (Bio-Rad) columns followed by ultrathin-layer PAGIF for the separation of the resulting simplified , enriched plasma fractions. The first fractionation step on Affi Gel Blue (Bio-Rad) columns further allows plasma to be fractionated in a variety of ways, since the elution profile may be altered simply by changing the elution conditions of buffer and temperature as shown by Gianazza and Arnaud (89). Some 27 fractions have been identified in the unbound, bound and strongly bound proteins by fused rocket immunoelectrophoresis as indicated in Fig. 7. The complexity of protein species present in a narrow pH range is illustrated in Fig. 8. This method of preliminary fractionation followed by further column fractionation also allows rapid purification of individual proteins as shown in Fig. 9 of the Group specific protein (Gc), which binds vitamin D_3 and actin on two different binding sites. The Gc can be separated readily from ceruloplasmin by Con-A. Presently some 85 alleles, of this highly polymorphic protein, have been identified.

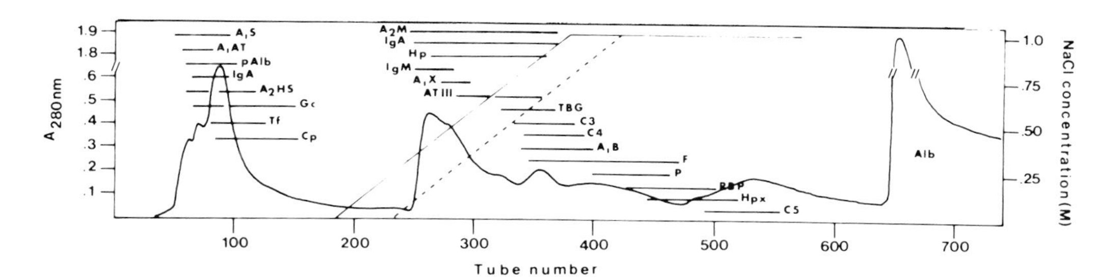

Fig. 7. Elution profile of plasma from Affi Gel Blue column at pH 7.2 and room temperatue. The column was equlibrated with 0.03 M phosphate buffer. (from Allen *et al.*) (90)

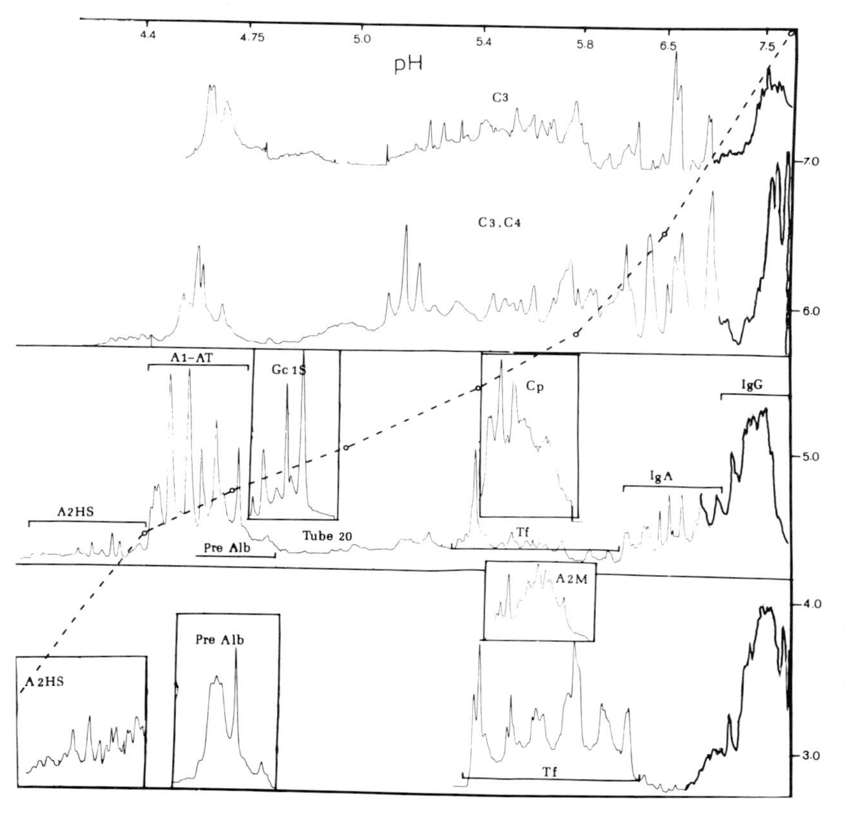

Fig. 8. Desitometric trace of the unbound fraction from a single 2 ml fraction from an Affi Gel Blue column separated on a pH 3 to 7 gradient. (from Allen and Arnaud) (25)

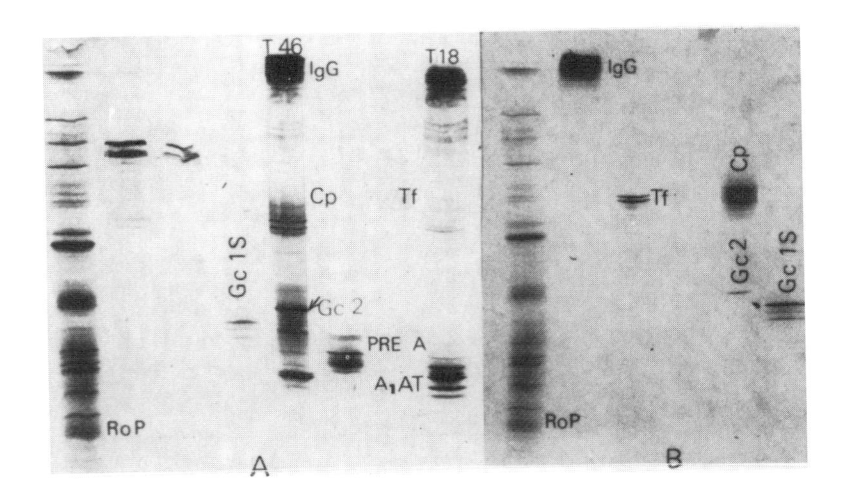

Fig. 9. Panel (A) Fraction number 46, known to contain IgG, ceruloplasmin (Cp), Gc and transferrin (Tf). Panel (B) Subfractionation of tube 46 on DEAE showing purification of Gc, IgG, Tf, Cp, Gc2 and Gc1S. (from Allen and Arnaud) (89)

4.2.4 Pseudo-ligand affinity chromatography followed by 2-D

Tracy *et al.* (20) have utilized Affi-Gel Blue (C. I. 60211) to remove albumin and IgG from serum prior to 2-D (ISO-DALT) separation to simplify the resultant pattern. The resulting maps may only represent the more abundant proteins in the plasma, with those minor proteins less than 50μ/dl unresolved or masked by the more concentrated proteins.Allen *et al.* (90,91) have subjected the fractions eluted from Affi-Gel Blue columns as described in the last section, concentated them by lyophilization and separated pooled fractions at a concentration of 1 mg/ml by the ISO-DALT 2-D procedure described by Anderson and Anderson (15,16) to determine both concentration effects on detection levels, potential overlap of spots that might occur in whole plasma and the efficacy of an initial simplification step on the polypeptide maps. The colored silver stain of Adams and Sammons (92) was used as an additional parameter to assist in the differentiation of spots on the 2-D maps. The following figures indicate selected fractions of the unbound, bound and strongly bound fractions eluted from the column by phosphate buffer, salt gradient and the chaotropic agent sodium thiocyanate respectively.

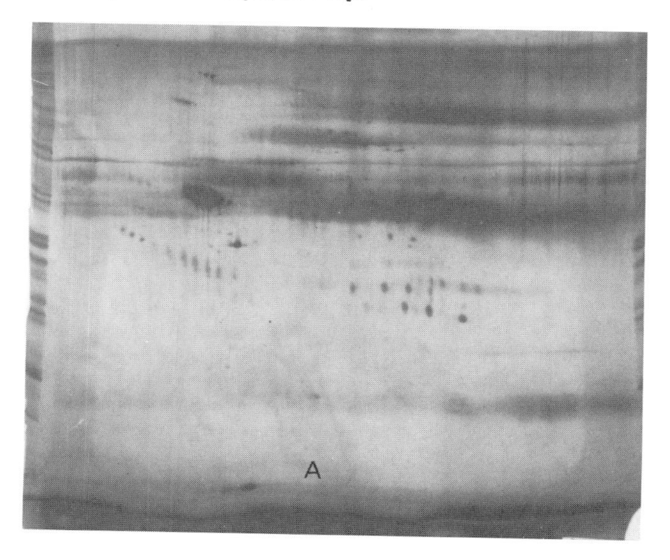

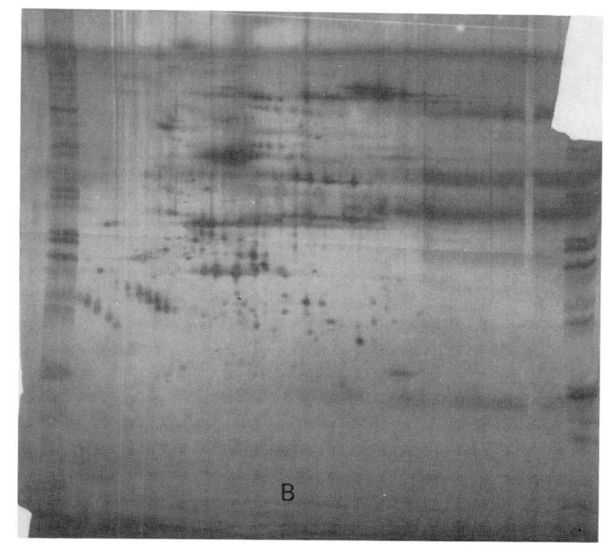

Fig. 10. Column eluates from 10 ml dialyzed plasma passed through an Affi-Gel Blue column 2.5 x 100 cm and collected in 4.8 ml fractions. Each ten tubes were pooled and lyophilized. Panel (A) represents tubes 30-39 eluted by 0.03 **M** phosphate buffer in the unbound fractions. Panel (B) is from tubes 300-309 of the bound plasma components eluted with a salt gradient. Gels were stained with silver by the method of Adams and Sammons (92) (from Allen *et al.*) (91)

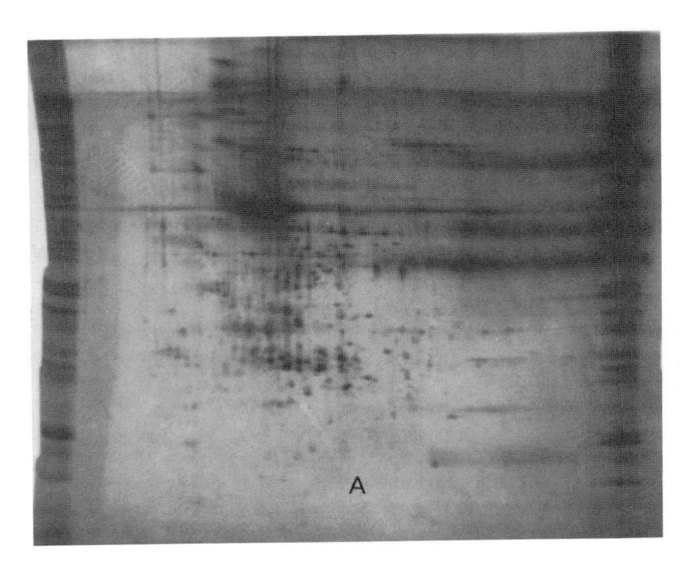

Fig. 11. Panel A, next series of tube from the salt gradient elution to panel (B) Fig. 10, tubes 310-319. Panel (B) Strongly bound fractions eluted with 0.5 **M** NaSCN (tubes 510-519). (91)

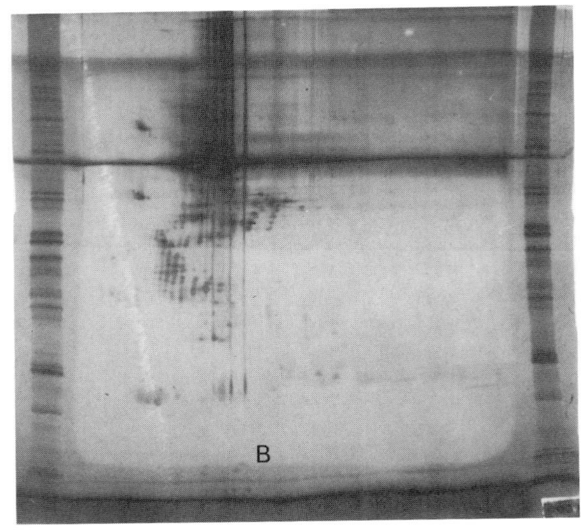

Selected composite drawings obtained from color enlargments of each gel are given in Figs. 12, 13 representing the unbound, bound and strongly bound plasma components.

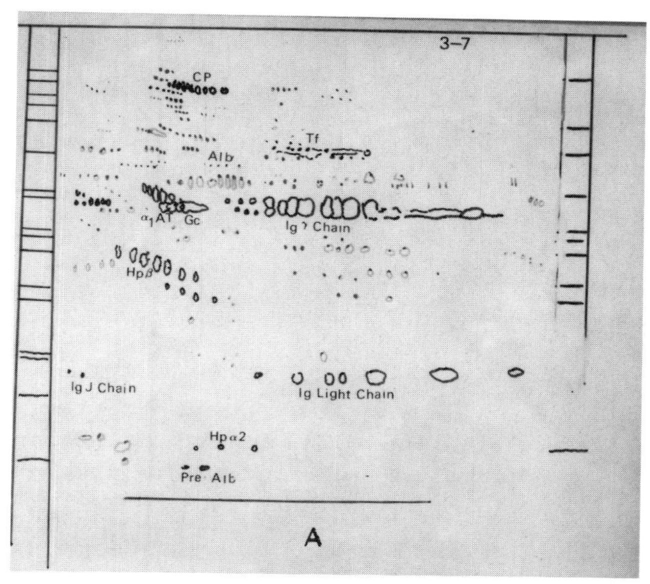

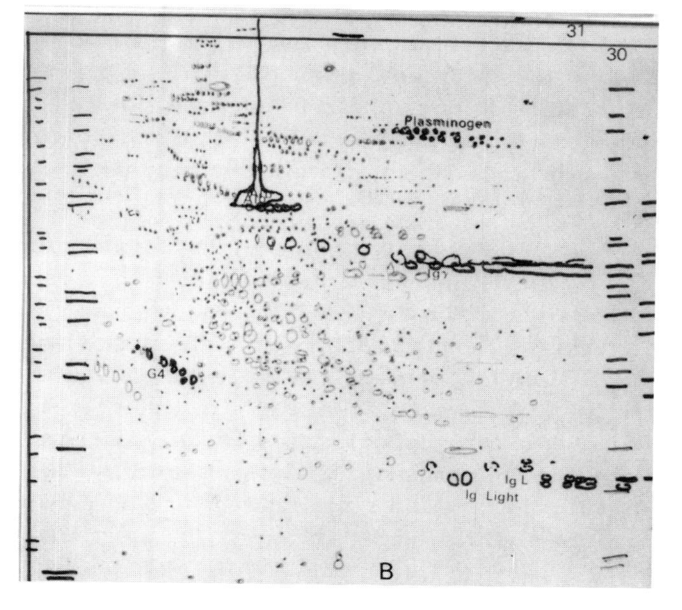

Fig. 12. Composite drawings from color photo enlargements. Panel (A) unbound fractions from tubes 30-79. Panel (B) tubes 300 - 319 from the salt gradient. (91)

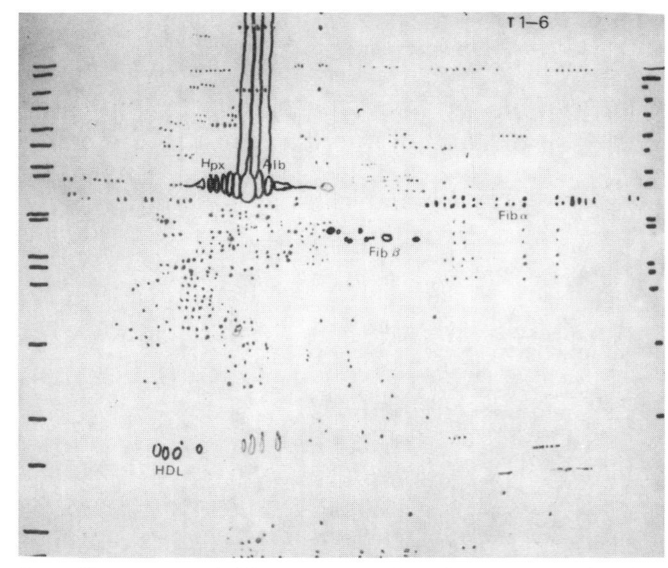

Fig. 13. Composite drawing of tubes 510-560 of the strongly bound plasma components eluted from the column with 0.5M NaSCN. (91)

In all cases of the composite drawings the lighter drawn spots are those that remain to be identified. Those protein spots identified were done so with the aid of the protein maps of Sammons *et al.* (93) and Anderson *et al.* (94) as well as with the corroboration of fused rocket results.

The exact mechanisms by which fractionation of plasma is accomplished using pseudo-ligand affinity chromatography on Affi-Gel Blue is not fully understood. It probably combines mostly hydrogen and ionic interactions and as performed in these examples can be considered to be due, partially, to ion exchange (90). In the plasma proteins, hydrophobic interactions do not appear to play a major role. Additionally each fraction in this method is enriched in its protein content, allowing a greater chance to detect minor components in concentrations of less than 25-50 μgm/dl.

O'Farrell (1) has postulated that some 10,000 polypeptide spots should be resolvable on 2-D gels from the observation that 100 bands can readily be resolved in each of the two dimensions. In the above examples it is readily apparent that present 2-D electrophoretic techniques alone probably can not, even with larger format gels, hope to resolve all of the proteins in a complex mixture such as plasma. This is apparent from the large number of overlapping spots found in only one-half of the gels examined to produce the composite drawings shown in Figs. 12, 13. While the composites shown in this study were not precisely controlled in each gel photo for slight swelling or shrinking, positions of spots on each gel may not be exactly as illustrated.However, this would not account for the great number of spots present in different fractions, which obviously overlap, were a total composite made of all of the fractions. This is readily illustrated in the individual pooled fractions from tubes 300-309 and 310-319 and in the composite of the pooled fractions in the salt gradient elution step in Figs 11,12. As many total spots are present as normally seen in a standard 2-D gel of whole plasma. In addition the use of lyophilization of each pooled fraction provides an effective concentration to 100 mg/dl protein allowing the detection of minor components that are in too low a concentration to detect by other means.

A further consideraton is that the 2-D electrophoresis, as performed above with gel rods used in the first dimension, may well have not resolved a number of the more acid or basic proteins as described by Dunn and Burghes (18). Thus, the actual total composite polypeptide map is probably considerably more complex than is indicated in the present literature.

It should be noted also that most of the enzymes which are present in the plasma possess NAD, NADP, or ATP as a coenzyme. Due to the pseudo-ligand effect of the immobilized dye used, these enzymes bind to the the dye and are eluted only by the chaotropic agent. Therefore, most of the additional protein families which are observed in these fractions are likley to correspond to such plasma components. With such techniques it is possible, without overloading the gel, to detect many more proteins and polypeptide spots. This is important, since it is a well established fact that for a number of proteins biological activity and control are accomplished with only minute amounts of the protein. Thus, multiparameter techniques have yet to be fully exploited in the detection and elucidation of protein interactions in biological phenomena, but offer an important adjunct to 2-D electrophoresis.

When a new protein or polypeptide is discovered by any of these procedures, that correlates consistantly with a given disease process, the technique has the built in capability also to be used for isolation

and structural identification of such spots. Subsequent production of specific antibody against newly found disease associated proteins can provide then specific reagents for routine screening, as suggested by Young and Tracy (34). Thus, it appears that the most important role of multiparameter separations and 2-D may be that of the initial detector. In the meantime, these higher resolution techniques provide a powerful "electromagnet" to locate the proverbial needle in the haystack, and in the case of biolgical applications many needles!

4.3. References

1. O'Farrell, P. H. : *J. Biol. Chem. 250,* 250 (1975).
2. Klose, J. : *Humangenetic 26,* 231 (1975).
3. Scheele, G. A. : *J. Biol. Chem. 250,* 5375 (1975).
4. Smithies, O. and Poulick, M. D. : *Nature* (London) *177,* 1033 (1956).
5. Margolis, J. and Kendrick, K. G. : *Nature* (London) *221,* 1056 (1969).
6. Dale, G. and Latner, A. L. : *Clin. Chem. Acta 24,* 61 (1969).
7. Macko, V. and Stegemann, H. : *Hoppe Seylers Z. Physiol. Chem. 350,* 917 (1969).
8. Weber, K. and Osborn, M. J. : *Biol. Chem. 244,* 4406 (1969).
9. Kaltschmidt, E. and Wittman, H. G. : *Anal. Biochem. 36,* 401 (1970).
10. Mets, L. S. and Bogorad, L. : *Anal. Biochem. 57,* 200 (1974).
11. Barret, T. and Gould, H. S. : *Biochem. Biophys. Acta 294,* 165 (1973).
12. Stegemann, H., Franksen, H. and Mako, V. Z. : *Naturforsch. 28,* 722 (1973).
13. Bahakdi, S, Knuferman, H. and Hoelz-Wallach, D. F. : *Biochem. Biophys. Acta 345,* 448 (1974).
14. Klose, J., Noah, J. and Kade, W. : in Radola, B. J., (ed.), *Electrophoresis '79* de Gruyter, Berlin, p. 297 (1980).
15. Anderson, N. G. and Anderson, N. L. : *Anal. Biochem. 85,* 331 (1979).
16. Anderson, N. L. and Anderson, N. G. : *Anal. Biochem. 85,* 341 (1979).
17. Anderson, N. G. and Anderson, N. L. : *J. Automated Chem. 2,* 177 (1980).
18. Dunn, M. J. and Burghes, A. H. M. : *Electrophoresis 4,* 97 (1983).
19. Dunn, M. J. and Burghes, A. H. M. : *Electrophoresis 4,* 173 (1983).

20. Tracy, R. P., Currie, R. M., Kyle, R. A. and Young, D. S. : *Clin. Chem. 28,* 900 (1982).

21. Willard, K. E., Giometti, C. S., Anderson, N. L., O'Conner, T. E. and Anderson, N. G. : *Anal. Biochem. 100,* 289 (1979).

22. O'Farrell, P. Z., Goodman, H. M. and O'Farrell, P. H. : *Cell 12,* 1133 (1977).

23. Burghes, A. H. M., Dunn, M. J., Statham, H. E. and Dubowitz, V. : *Electrophoresis 3,* 185 (1982).

24. Allen, R. C. :*Electrophoresis 1,* 32 (1980).

25. Allen, R. C. and Arnaud P. : *Electrophoresis 4,* 205 (1983)

26. Dunn, M. J., Burghes, A. H. M., Thompson, B. J., Statham, H. E. and Dubowitz, V. : in Stathakos, D. (ed.), *Electrophoresis '82,* de Gruyter, Berlin, p. 641 (1982).

27. Allen, R. C., Christopher, J., Lorincz, L., Allen, R. C. Jr. and Liu, P. : in Radola, B. J. (ed.), *Elektrophorese Forum '80,* Technische Universität München, München, p. 117 (1980).

28. Thompson, B. J., Dunn, M. J., Byrghes, A. H. M. and Dubowitz, V. : *Electrophoresis 3,* 307 (1982).

29. Burdett, P. and Lizana, J. : in Allen, R. C. and Arnaud, P. (eds.), *Electrophoresis '81,* deGruyter, Berlin, p. 329 (1981).

30. Burghes, A. H. M., Dunn, M. J. and Dubowitz, V. : *Electrophoresis 3,* 354 (1982).

31. Görg, A., Postel, W., Westermeier, R., Gianazza, E. and Righetti, P. G. : in Allen, R. C. and Arnaud, P. (eds.), *Electrophoresis '81* , de Gruyter, Berlin, p. 257 (1981).

32. Kinzkofer,a. and Radola, B. J. : *Electrophoresis 4,* 408 (1983).

33. Tracy, R. P., Currie, R. M. and Young, D. S. :*Clin. Chem. 28,* 908 (1982).

34. Young, D. S. and Tracy, R. P. : *Electrophoresis 4,* 117 (1983).

35. Wykoff, M., Rodbard, D. and Chrambach, A. : *Anal. Biochem. 78,* 459 (1977).

36. Rothe, G. M. and Purkhandaba, H. : *Electrophoresis 3,* 33 (1982).

37. Rothe, G. M. and Purkhandaba, H. : *Electrophoresis 3,* 43 (1982).

38. Narayanan, K. R. and Raj, A. S. : in Radola, B. J. and Graesslin, D. (eds.), *Electrophoresis and Isotachophoresis* , de Gruyter, Berlin, p. 221 (1977).

39. Pretch, W., Charles, D. J. and Narayanan, K. R. : *Electrophoresis 3,* 142 (1982).

40. Klose, J. and Feller, H. : *Electrophoresis 2,* 12 (1981).

41. Horst, M. N., Mahaboob, S., Basha, M., Baumbach, G. A., Mansfield, E. H. and Roberts, R. M. : *Anal. Biochem. 102,* 309 (1980).

42. Wilson, D. L., Hall, M. E., Shone, G. C. and Rubin, R. W. : *Anal. Biochem. 83,* 33 (1977).

43. Stegemann, H. : in Righetti, P. G., Van Oss, C. J. and Vanderhoff, J. W. (eds.), *Electrokinetic Separation Methods*, Elsivier, Amsterdam, p. 313 (1980).

44. Ames, G. F. L. and Nikaido, K : *Biochem. 15,* 616 (1976)

45. Harrell, D. and Morrison, M. : *Arch. Biochem. Biophys. 171,* 161 (1979).

46. Tanford, C. E. : *The Hydrphobic Effect&,* 2nd Edition, Wiley, London, (1980).

47. Beacker, P. : in Schick, M. J. (ed.), *Non-ionic Surfactants* , Arnold, London, p. 478 (1967).

48. Murkegee, P. : in Mittal, K. L. (ed.), *Micellization, Solubilization and Microemulsions* , Plenum Press, New York, p. 191 (1977).

49. Allen, J. C. and Humphries, C. : in Aburthnot, J. P. and Beckley, J. A. (eds.), *Isoelectric Focusing* , Butterworth, London, p. 347 (1975).

50. Goenne, A. and Ernst, R. : *Anal. Biochem. 87* , 28 (1978).

51. Helmeland, L. M., Nebert, D. W. and Chrambach, A. : *Anal. Biochem. 95,* 201 (1979).

52. Merril, C. R., Goldman, D. and Ebert, M. : in Allen, R. C. and Arnaud, P. (eds.), *Electrophoresis '81,* deGruyter, Berlin, P. 343 (1981).

53. Bustos, S. E. and Fung, L. : in Allen, R. C. and Arnaud, P. (eds.), *Electrophoresis '81,* deGruyter, Berlin, P. 316 (1981).

54. Singh, S., Wllers, I., Goedde, H. W. and Klose, J. : in Allen, R. C. and Arnaud, P. (eds.), *Electrophoresis '81,* deGruyter, Berlin, P. 289 (1981).

55. Thorsrud, A. K., Vatn, M. H. and Jellum, E. : *Clin. Chem. 28,* 884 (1982).

56. Tollaksen, S. L., Anderson, N. L. and Anderson, N. G. : *Argonne National Laboratory* BIM-81-1 (1981).

57. Tracy, R. P., Currie, R. M. and Young, D. S. : *Clin. Chem. 28,* 890 (1982)

58. Jackle, H. : *Anal. Biochem. 98,* 81 (1979).

59. Radola, B. J. : *Electrophoresis 1,* 43 (1980).

60. Bjellqvist,, B., Ek, K., Righetti, P. G., Gianazza, E., Görg, A. and Westermeier, R. : *J. Biochem. Biophys. Meth. 6* , 317 (1982).

61. Chait, E. A. Personal Communication.

62. Altland, K. and Kaempfer, M. : *Electrophoresis 1,* 57 (1980).

63. Merril, C. R., Goldman, D. and Van Keuren, M. L. : *Electrophoresis 3,* 17 (1982).

64. Taylor, J., Anderson, N. L. and Anderson, N. G. : *Electrophoresis 4,* 338 (1983).

65. Merril, C. R., Goldman ,D. and Ebert, M. : in Allen, R. C. and Arnaud, P. (eds.), *Electrophoresis '81,* de Gruyter, Berlin, p. 343 (1981).

66. Skolnick, M. M., Stanley, R, Sternberg, S. R. and Neal, J. V. : *Clin. Chem. 28*, 969 (1982).

67. Taylor, J., Anderson, N. L. and Anderson, N. G. : in Allen, R. C. and Arnaud, P. (eds.), *Electrophoresis '81*, de Gruyter, Berlin, p. 383 (1981).

68. Lemkin, P. F. and Lipkin, L. E. : in Allen, R. C. and Arnaud, P. (eds.), *Electrophoresis '81*, de Gruyter, Berlin, p. 401 (1981).

69. Lemkin, P. F. : *Electrophoresis 4*, 71, (1983).

70. Skolnick, M. M. : *Clin. Chem. 28*, 979 (1982).

71. Van Keuren, M. L., Goldman, D. and Merril, C. R. : in Allen, R, C. and Arnaud, P. (eds.), *Elecrophoresis '81*, de Gruyter, Berlin, p. 355 (1981).

72. Comings, D. E. : *Clin. Chem. 28*, 782 (1982).

73. Comings, D. E., Carraway, N. G. and Pekkula-Flagan, A. : *Clin. Chem. 28*, 790 (1982).

74. Comings, D. E. : *Clin. Chem. 28*, 805 (1982).

75. Comings, D. E. and Pekkula-Flagan, A. : *Clin. Chem. 28*, 813, (1982).

76. Jellum, E. and Thorsrud, A. K. : *Clin. Chem. 28*, 876 (1982).

77. Edwards, J. J., Tollaksen, S. L. and Anderson, N. G. : *Clin. Chem. 28*, 941, (1982).

78. Bravo, R. and Celis, J. E. : *Clin. Chem. 28*, 949 (1982).

79. Lundberg-Holm, K., Bagley, E. A., Nusbacher, J. and Heal, J. M. : *Clin. Chem. 28*, 962 (1982).

80. Hanash, S. M., Tubergen, D. G., Heyn, R. M., Neel, J. V., Sandy, L., Stevens, G. S., Rosenblum, B. B. and Krzesiki, R. F. : *Clin. Chem. 28*, 1026 (1982).

81 Willard, K. E. : *Clin. Chem. 28*, 1031 (1982).

82. Scheele, G. A. : *Clin. Chem. 28*, 1056 (1982).

83. Altland, K. and Hackler, R. : in Radola, B. J. (ed.), *Elektrophorese Forum '80* Technische Universität München, München, 22 (1980).

84. Altland, K., Röder, Th, Jakin, H. M., Zimmer, H.-G. and Neuhoff, V. : *Clin. Chem. 28*, 1000 (1982).

85. Dermer, G. B. and Edwards, J. J. : *Electrophoresis 4*, 212 (1983).

86. Pearson, T. and Anderson, N. L. : *Anal. Biochem. 101*, 377 (1980).

87. Arnaud, P., Galbraith, R. M., Chapuis-Cellier, C., Galbraith, G. M. P. and Fudenberg, H. H. : in Peeters, H., (ed.), *Protides of the Biological Fluids*, Proc. 26th Colloquium, Pergamon Press, New York, p 649 (1979).

88. Allen, R. C. and Arnaud, P. : in Allen, R. C. and Arnaud, P. (eds.) *Electrophoresis '81*, de Gruyter, Berlin, p. 167 (1981).

89. Gianazza, E. and Arnaud, P. : *Biochem. J. 203*, 637 (1982).

90. Allen, R. C. Arnaud, P., Sammons, D. W. and Adams, L. D. : in Hirai, H. (ed.), *Electrophoresis '83*, de Gruyter, Berlin, In Press.

91. Allen, R. C., Arnaud, P., Sammons, D. W. and Adams, L. D. : in Radola, B. J. (ed.), *Elektrophorese Forum '83*, Bode, München, p. 30 (1983).

92. Adams, L. D. and Sammons, D. W. : in Allen, R. C. and Arnaud, P. (eds.), *Electrophoresis '81* de Gruyter, Berlin, p. 155 (1981).

93. Sammons, D. W., Adams, L. D. and Nishizawa, E. E. : *Electrophoresis 2*, 155 (1981).

94. Anderson, N. L., Nance, S. L., Pearson, T. W. and Anderson, N. G. : *Electrophoresis 3*, 135 (1982).

5. COMPONENT VISUALIZATION

5.1. Protein Staining

5.1.1. Background

Since the introduction of high resolution electrophoresis in polyacrylamide gel by Raymond and Weintraub (1) and the introduction of the Disc electrophoresis system by Ornstein and Davis (2), there has been a tremendous amount of effort in the search of suitable protein, glycoprotein and lipoprotein stains. Busse (3) in 1969 clearly defined the requirement of an ideal protein stain, which should have the following characteristics:

1. The stain should stain all the different proteins quickly and with the same intensity.
2. It should quickly and completely be able to be washed out of the support material.
3. It should have a linear and stoichiometric relationship between protein binding and protein content.

The development of even higher resolution methods and difficulties with conventional dyes has led to the search for more sensitive and specific staining procedures and the exacting requirements of Busse have been more or less abandomed in light of practical considerations. The earlier stains, such as Amido Black, Ponceau S and Lissamine green have been largely replaced with the Coomassie series of dyes which are more sensitive and can be used effectively for the staining of PAGIF and AGIF. Their range of linearity is over a 20 to 30 fold concentration difference, with a detection capability of proteins in plasma at a level of 10 mg/dl and a sensitivity level of 50 to 100 ng/mm^2 (4).

Protein dyes were reviewed extensively in the first edition of this book, many of which are no longer in use. However, the more common of these and their sensitivity relative to Coomassie Brilliant Blue R250 are included in Table 1. The various fixatives that were commonly employed in these earlier studies, have, in light of recent studies on ultrathin–layer gels by Radola (4), been shown to be inappropriate to obtain maximal sensitivity. He has demonstated that acid–alcohol fixatives really do not fix the protein in the gel which causes leaching of the protein during subsequent staining and destaining. Thus, today 20 per cent trichloroacetic acid appears to be the fixative of choice for proteins, and as Saravis (6) has reported, glutaraldehyde fixation of the smaller polypeptides is necessary also, to prevent their leaching during stain procedures, particularly with immuno–enhancement methods.

Grubhofer (7) has reviewed the dyes that may be used in both PAGE and PAGIF. These are shown in Table 1., which compares the protein adsorption of the dye in column 1 with the ease of destaining in column 2, as compared to Coomassie Brilliant Blue R–250. Thus, a dye with a protein adsorption of greater than 1.0 is desirable, as is one whose destaining value is less than 1.0. The sensitivity of each is given in column 3.

Table 1. Staining and destaining characteristics of various dyes used in protein staining.

Stain	Absorption	Destaining	Sensitivity
Coomassie Brilliant Blue R-250	1.0	1.0	50 ng/mm²
Coomassie Brilliant Blue G-250	0.8	0.4	60 ng/mm²
SERVA Violet 49	0.4	0.07	125 ng/mm²
Amido Black 10 B	2.8	1.5	150 + ng/mm²
Procion Brilliant Blue RS	0.2	0.4	150 ng/mm²
Ponceau S	2.1	2.45	200 ng/mm²
Acid Fuchsin	1.97	0.96	150 + ng/mm²
Lissamine Green	0.98	0.47	150 + ng/mm²

5.1.2. Procedures

There have been a number of suggested procedures for staining PAGE, SDS–PAGE and PAGIF gels with Coomassie Brilliant Blue R–250. The most rapid are those using staining at elevated temperatures as given below:

Table 2. Rapid Coomassie Brilliant Blue R–250 staining 100μ to 3mm thick gels, (from Maurer and Allen) (8).

1. **Fixation**	20 % TCA 10 min to 1 hour.
2. **Stain solution**	2 % Coomassie Brilliant Blue R–250 4.0g Coomassie Brilliant Blue R–250 950 ml 95 per cent undenatured alcohol 850 ml distilled water 200 ml Glacial acetic acid Dissolve dye in alcohol, add water and acid and filter.
3. **Procedure**	Stain (125μ for 1 min, 1.0 mm for 10 min and 3.0 mm for 30–45 min at 50 –60 °C in a shaking water bath.
4. **Destain solution**	2100 ml 95 per cent alcohol 800 ml Glacial acetic acid 5100 ml Distilled water.
5. **Destain procedure**	Destain in solution 4 at 50 – 60 °C changing destain solution when it reaches a deep blue color until clear background results. (three to 40 min. depending on gel thickness). The color of the bands may be enhanced by placing the destained gel in 10 per cent acetic acid at 55 °C for several minutes.

5.2. Glycoprotein Stains

5.2.1. PAS stain

A number of histological methods have been employed for the specific staining of the glycoproteins. A number of these are based on the periodic acid Schiff stain (PAS) for vicinal glycol rich glycoproteins developed by McManus and Hoch-Ligeti (9). An adaptation of this technique for acrylamide gel electrophoresis by Allen *et al.* (10) is given below.

Table 3. Modified PAS stain for PAGE and PAGIF

1. **Fixation**	20 per cent TCA for 10 to 30 min depending on gel thickness (See Coomassie Blue procedure).
2. **Wash**	Wash the fixed gel for several minutes with running distilled water.
3. **Oxidation**	Place in 1.0 per cent periodic acid solution for 20 min. (1.0 mm gel, increase or decrease depending on gel thickness)
4. **Staining**	The periodic acid solution is decanted and a mixture of 1 part 1.0 per cent periodic acid to 5 parts of Schiff base (Fisher Scientific) is added for 10 min. The gel will turn dark brown. Then add fresh Schiff base to the gel which will rapidly clear with the PAS positive bands staining a bright red.

As a note of caution the Schiff base must be fresh and should be stored at refrigerator temperature and not used after the expiration date on the bottle. Each bottle should be tested for activity prior to use.

5.2.2. Specific Lectin Probes

The PAS reaction for glycoprotein staining is a general one and gives little indication of the type of carbohydrate moieties present. Further information may be obtained in identifying those carbohydrates present by utilizing specific lectin probes, such as Concanavalin–A; specific for internal or terminal mannose in the α configuration and N-acetylglucosamine in the terminal position. This lectin is also specific for 1–4 α linked glucose, which occurs rarely, if ever, in glycoproteins (11). A general method for lectin probes is given in the following example of the Concanavalin–A – horse radish peroxidase stain.

Table 4. Con–A –HRP stain

1. **Fixation**	A.	20 per cent TCA for 10 min. (125μ to 1.0mm gels)
2. **Reagents**	A.	5.0 mg Con–A (Sigma grade IV;salt–free) dissolved in 50 ml of PBS at pH 7.2.
	B.	1.0 mg horse radish peroxidase in 100 ml of above PBS.
	C.	Diaminobenzidine. 6.0 mg dissolved in 0.05M Tris.Cl (0.61 g Tris/100 ml adjusted to pH 7.6 with 1.0 N HCl). Just before use add 6 drops H_2O_2.
3. **Procedure**	A.	The fixed gel is washed twice with PBS for 2 min each wash.
	B.	Then react the gel with the Con–A–HRP solution for 90 min. at room temperature with constant agitation.
	C.	The gel is then washed 3 times for 5 min each in PBS.
	D.	React gel with DAB–peroxide until bands reach desired intensity.

Additional enhancement of the peroxidase stain may be achieved by reacting the gel in 1.0 per cent osmium tetroxide for 5 min

The three stain procedures described above are illustrated in Fig. 1.

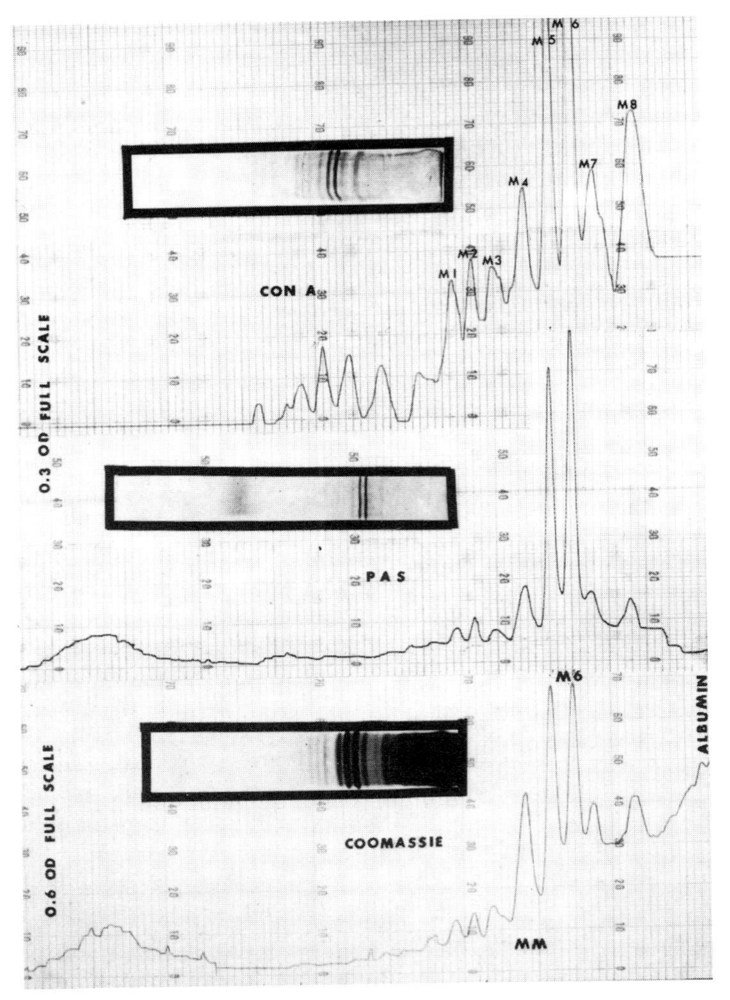

Fig. 1. Pi type MM serum separated by PAGIF on a 1.0mm slab with LKB Ampholine on a pH 3.5 to 5.0 gradient, comparing Con-A - HRP, PAS and Coomassie Brilliant Blue R-250. Densitometric traces of each separation are aligned with the M_6 band. (from Allen *et al.*) (10)

5.2.3. Other glycoprotein stains

Acid glycoproteins can be stained by incubation for extended periods, depending on gel thickness, in 0.2 per cent Alcian Blue, Caldwell and Pigan (12). Acid mucopolysaccharides may be stained in 1.0 per cent Toluidine blue in 3.5 per cent acetic acid as described by Rennert (13).

5.3. Silver Staining

5.3.1. Background

In the last four years there has been an increasing use of silver to stain proteins, polypeptides, nucleic acids and lipoproteins with either free ionic silver, or in the form of diammine complexes. These methods have been developed from silver—protein complex studies done as early as 1880 (14) to the first practical application of silver diammine as a histochemical stain for degenerated nerve tissue described by Nauta and Gygax (15) in 1951. The great advantage of this technique is that reported sensitivity is 20—200 times more senstive than the best previously available protein stains such as Commassie Brilliant Blue. Detection levels of 30—100 picograms protein per mm^2 are now possible compared to 10ng per mm^2 with previous methods.

The technique has evolved into two general procedures. The first utilizes a diammine silver stain following treatment with an aldehyde and the second uses uncomplexed silver from silver nitrate. Both procedures are widely used, however, there has been a proliferation of modifications in each. The chemistry of both staining reactions has received considerable attention and speculation in the literature. However, at present, the precise chemical mechanisms involved in each method remain unclear with each variation arrived at more or less empirically, further confusing the elucidation of reaction mechanisms. Two points appear quite clear in all procedures. The first is the necessity to assure that the acid proteins do not remain protonated during the reaction or reduction procedure, otherwise an incomplete reaction with weak staining of such proteins results. The second is purity of reagents, particularly the need for deionized or double distilled water, although one report describes the necessity to use impure alcohol, containing aldehydes, if the normal procedure does not work (16).

5.3.2. Steps in the various procedures

5.3.2.1. Fixation of proteins following electrophoresis

One of the critical considerations in effectively utilizing the sensitive silver staining methods is the complete fixation of the proteins following isoelectric focusing in SDS-PAGE, or in two-dimensional procedures. Various methods have been reported for protein fixation, although two general procedures are employed; acid-alcohol and trichloroacetic acid. The former fixative, in ultrathin-layer gels (thickness 50 to 250μ) will not fully fix the proteins leading to protein loss by leaching. Thus, trichloroacetic acid at a concentration of 20 percent is recommended especially for gels less than 0.5mm thick (4,17). Sammons, *et al.* (18) have shown also, that if colored silver staining is required in thicker 2-D gels, that ethanol rather than methanol is necessary. Adequate fixation is a prerequisite if one is to take full advantage of the sensitivity of silver stains, particularly in high resolution ultrathin-layer isoelectric focusing techniques.

The fixation acid can also affect color reproducibility as applied to two-dimensional gels. When nitric acid is substituted for acetic in 2-D gel fixation, haptoglobin shows no color change. However, the fibrinogen α-chain changes from a yellow color, following the use of ethanol-acetic acid fixation and to a red color following the use of ethanol-nitric acid fixation (18).

5.3.2.2. Diammine silver stains

In the diammine stains initial treatment with paraformaldehyde, glutaraldehyde or glutardialdehyde is utilized. While the aldehydes may act as mild oxidants, they more probably cross link with the proteins providing additional free aldehyde groups with which to subsequently bind additional silver and works best at an alkaline pH of 7.2 to 7.4. Saravis (6) has reported that the direct fixation in glutaraldehyde fumes works exceedingly well to prevent the loss of small peptides that can be washed from the gel, even with 20 per cent TCA . Marshall and Latner (19) have reported that when the stain procedure employs ammonia-diammine that glutaraldyhyde is the agent of choice, producing the most sensitive subsequent staining. But, when methylamine is substituted for ammonia in the diammine preparation, they have found paraformaldehyde performs best.

In the diammine procedures employing aldehydes, complete washing out of the aldehyde is a most critical step in obtaining a subsequent stain free of background. This may be aided by reducing the glutaraldehyde concentration from 4.5 to 2.2 percent and reacting the fixed proteins at 50 °C for 20 min. in 1.5mm thick gels 1.5 min. in 125μ gels, which reduces required wash times. However, in gels over 1mm thick an initial hour of washing at 50 °C followed by an overnight wash is required to remove excess unbound glutaraldehyde, which could subsequently bind silver yielding undesirable background. Without complete removal of the glutaraldehyde an orange to reddish–orange background will occur during or following the silver reduction step. Gels affixed to glass or polyester sheets with silane are particularly prone to background staining of the gel at the support interface if insufficiently washed, or if reacted too long in the silver diammine. Thus, in procedures using recycling to enhance stain intensity, short recycling times are best for optimal results.

The diammine procedure initially used rather high silver concentrations to saturate the proteins which was a disadvantage from a cost standpoint. In more recent methods, the silver concentration has been reduced by a factor of eight to fifty–fold while, retaining good sensitivity. The amount of silver used in the various procedures to actually stain the protein is similar and the unbound silver can be recovered by dumping the silver containing reaction mixture into 10 per cent HCl and recovering the silver as insoluble AgCl.

A further point of consideration in the diammine procedure is the ammonium/sodium ratio. Nauta and Gygax (15) first described this effect in reference to stain specificity. They found that higher ratios were most suitable for the histological staining of degenerated nervous tissue, where in retrospect only the glycoproteins were desired to be stained. A ratio of 10.5:1 stained glycoproteins specifically while a ratio of 4.5:1 stained all nervous tissue protein non–specifically. With fixed proteins following electrophoretic separation, Schwitzer, *et al.* (20) utilized a ratio of 28:1 but it was found that while α_1– antitrypsin, a glycoprotein, stained well, the albumin stained a light brown with this method. Decreasing the ratio progressively from 16:1 to a ratio of 8.0:1 or lower produced better non-specificity as evidenced by a monochromatic black stain of the albumin at lower ratios with a sensitivity of 0.1 ng protein mm^2 (21). These latter ratios have been found suitable for a number of biological fluids, cell and bacterial extracts. Guevera, *et al.* (22), on the other hand, have employed a ratio of approximately 4:1 with a resulting protein sensitivity of less than 0.03 ng/mm^2.

While the diammine stains do not produce the more vivid colors of the non–diammine ionic silver methods, various serum proteins do have distinct colors. For example, α_1–antitrypsin is blue–black and orosomucoid yellow to yellow–orange, and a number of other serum proteins give consistant, reproducible, colors aiding further in their identification (23).

Gels previously stained with Coomassie Brilliant Blue R250 ranging from 100 μ thick to 1.5mm thick may be readily counterstained with silver following a 5 min to 1 hour wash in 10 per cent alcohol at 50 °C in a shaking water bath. Gels up to two years old, both dried as films and stored in acid, may be successfully counterstained with the diammine method. However, should there be a question of original fixation procedure, refixing in 20 percent TCA is probably advisable (23).

5.3.2.3. Staining with ionic silver

A mild oxidation step rather than aldehyde treatment is used in a number of non–diammine methods as recently reported by Merril, *et al* . (17), Ansorge, for protein (24) and Guillemette and Lewis (25) and Goldman, *et al* . (26) for nucleic acids. Whether this serves as a photo reversal step as suggested by Merril (17), or actually produces free aldehydes is not at present clear. Ansorge (24) suggests that oxidation alone is not the critical step, but that the appropriate metal ion is the important factor. He reported poorer results with dichromate in regard to sensitivity than with permanganate and copper. On the other hand, the method of Adams and Sammons (27), for obtaining a polychromatic stain has no oxidation step, or copper enhancement involved, yet the sensitivity is essentially equivalent.

When potassium dichromate–nitric acid oxidation is used, reaction times of 5–30 min only are required on 0.8mm and thicker gels and 5 min or less, on gels under 0.5mm thick. Very short wash times are required, *i.e* . four changes over a 1 min period for 0.8mm thick gels. Reaction times range from 5–30 min for 0.8mm gels to two hours or more in the Adams and Sammons (27) method for 1.5mm Iso–Dalt gels. These procedures employing oxidation first by, *i.e.* potassium dichromate–nitric acid, potassium permanganate, or no oxidation step, all use silver nitrate directly, usually at a level of 0.2 per cent which is considerably less than the diammine procedures.

Enhancement of sensitivity by recycling in the silver, either with the diammine procedure, as previously mentioned, or with the quicker ionic silver procedures is equally effective. This procedure effectively intensifies bands; however, if not carefully carried out, may result in a heavy background. This usually occurs where the recycling was carried out for too long a time. Dark backgrounds will then require a controlled lightening procedure which also may reduce band intensity and thus, the gains of recycling may be lost. On the other hand, gels may be only *primed* by the initial silver reaction step with very weakly stained or no bands visible initially. In such cases, a second silver reaction step is necessary to get the bands to stain at their maximal intensity. In such instances background development following a longer recycling time is not usually a problem.

5.3.2.4. Development of silver by reduction to its metallic form

There are three commonly employed procedures for the reduction of ionic silver to metallic silver. The first is an acid method employing citric acid and formalin first described by Nauta and Gygax (15), the second is a basic procedure employing sodium carbonate and formalin and the third, also a basic procedure, with sodium hydroxide and formalin. The last procedure produces the most vivid colors. The acidic system in the author's experience, while slower, is easier to control. Heavy surface silvering is not a problem and reduction of fully silver–saturated proteins may be obtained without any appreciable background. Background often can be avoided, by stopping the reduction reaction with 10 percent acetic acid and then washing in water, followed by lightly rubbing the surface with a wet cotton swab. When the cotton no longer evidences reduced silver, the gel may be washed again with water and then dried. This procedure produces a marked reduction in what was often presumably considered background; but was in actuality a result of an invisible film of silver remaining on the gel surface (23). The more rapid, basic procedures (27) can quickly deposit silver on the surface of the gels even though they have been reacted initially with six-fold lower silver concentrations. A subjective opinion is that the basic procedure is more sensitive; but is harder to control in regard to background, particularly when utilized with ultrathin-gels in combination with the diammine technique. With basic reduction procedures a yellow to yellow-orange background is normally present, although if evenly distributed, causes no problem in photography or subsequent semiquantitative microdensitometry. The choice of reduction method rests mainly on each investigator's requirement and preference.

The third procedure (27) employs a very basic system for polychromatic staining. Here sodium hydroxide – formalin reduction is followed by sodium carbonate enhancement of the colored proteins. Again, a yellow to reddish background is apparent in gels from 250μ to 1.5mm thick stained completely through the protein bands, or polypeptide spots in 2-D electrophoresis. This procedure is extremely rapid on ultrathin-layer gels which offers an additional advantage of specifically colored protein zones. Consideration may be given to using a somewhat higher concentration of silver nitrate in the initial reaction step with subsequently a more controlled development time through reduction in the amount of formalin. It is also perhaps more efficacious to utilize a moderately increased sample concentration where practical, rather than trying to push a particular silver staining procedure to its limit of sensitivity with the risk of an increased background (see Fig. 2.).

5.3.2.5. Reaction temperatures

Many authors have been quick to point out the necessity of maintaining a constant temperature for each reaction. Recommmendations range from 20 to 60 ºC from various authors. It should be taken into consideration that ambient temperatures throughout the world flucuate widely during the year and that all laboratories are not air conditoned or well temperature controlled, suffering wide variations in temperature during both winter and summer months. An ambient temperature shift of 3–5 °C can markedly effect staining results, particularly with diammine procedures. Marshall and Latner (19) utilized 60 °C for the first steps in their procedure with excellent results to overcome this difficulty, while others have used 20 °C.

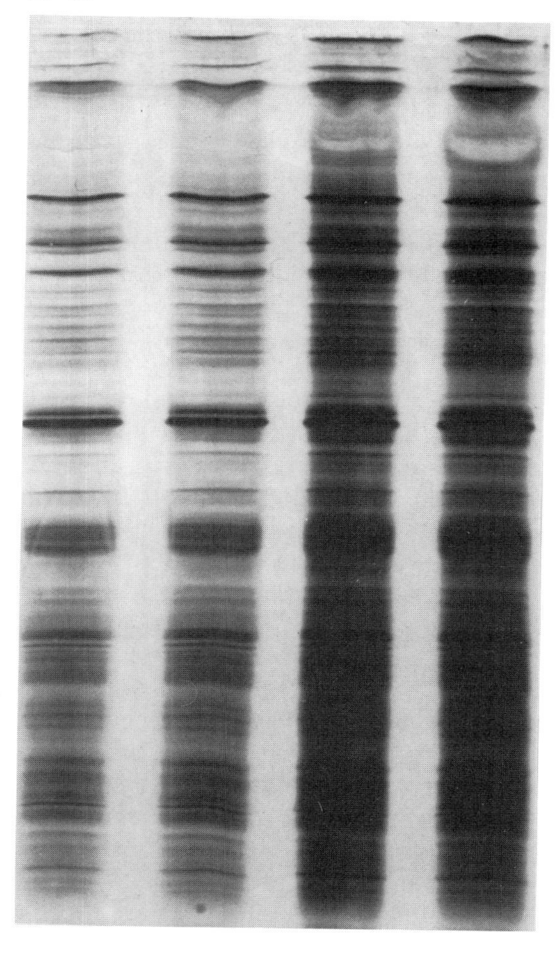

Fig. 2. Diammine silver stain of Rohament P™ enzyme preparation separated by ultrathin – layer PAGIF using Servalyte™ pH 3 to 7 gradient. Separations: (A) 0.5 µl of a 0.1 per cent protein concentration, (B) 1.0 µl of (A), (C) 0.5 µl of a 1.0 per cent protein concentration, (D) 1.0 µl of the same concentration as in (C). (From Allen) (28)

Utilizing 50 °C in all steps, but the final reduction in the diammine procedure, improves year round reproducibility. Staining in a constant temperature water bath is therefore a ready approach to obtain day to day and season to season reproducibility in many of the silver staining procedures (23).

5.3.2.6. Staining proteins on agar gel

There have been numerous attempts to use the silver stain on agarose as a supporting media without success either with free ionic silver or with diammine complex methods. Recently Willoughby and Lambert (29) have reported a room temperature stain carried out in a single staining and reduction step where the protein is stained in the presence of a silver–tungstosilicic–acid–ammonium complex. Their procedure is reportedly 10 times as sensitive as Coomassie Brilliant Blue as compared to the 100–200 fold increase for the silver procedures described on acrylamide gels . The chemistry of this procedure is not clear and a pH of 11 to 12 is necessary. The tungstosilicic–acid appears only to control the reduction rate of the silver ions in the staining complex. Budowle (30) has reported that Coomassie Brilliant Blue R–250 staining first and then counterstaining with this procedure provides increased sensitivity and reduced background staining.

5.3.2.7. Sample concentration

It should be recognized that samples such as serum have a wide range of individual serum component concentration. For example, there are 43μg of albumin per μl and only 0.1 μg per μl of total orosomucoid. In the latter protein with 8–9 multiple molecular forms present, this gives a variation in serum protein concentration of the individual components of up to 100 fold. At present, the range of linearity of this method, (some 30–40 fold), does not allow accurate interprotein band quantification and comparison to the degree presently accepted with Coomassie Brilliant Blue R250 (31). However, comparisons of the same protein between samples may readily be carried out with reasonable quantitative accuracy.

5.3.2.8. Other Applications

Tsai and Frausch (32) have reported a modification of the diammine silver technique using periodic acid oxidation followed by diammine silver for the staining of lipopolysaccharides separated by electrophoresis in polyacrylamide gels. A brown stain is produced in the lipopolysaccharide containing bands. Interestingly, similar treatment of serum proteins with periodic acid initially, followed by diammine silver staining, does not appear to be an effective method staining either proteins or glycoproteins.

5.4. Silver Staining Methods

Table 5. Diammine silver in 1mm thick gels (from Oakley *et al.*) (33)

1. **Fixation**	10 per cent gluteraldehyde or ethanol : 15 per cent acetic acid
2. **Staining solution**	2.8 ml of NH_4OH (concentrated stock solution)
	42.0 ml of 0.36 percent NaOH solution
	8.0 ml of 20 per cent $AgNO_3$, add water to bring volume to 200 ml.
3. **Developer.**	0.05 g/liter citric acid to which is added 0.5 ml of 37 per cent formaldehyde
4. **Destaining solution.**	25 per cent solution of Kodak Rapid Fixer A.
5. **Destaining stop solution.**	25 per cent Kodak Hypo Clearing Agent.
	STAINING
1.	Place gel in fixing solution for 45 min
2.	Rinse the gel in 750 ml of double distilled H_2O for 10 min with agitation.
3.	Add 100ml stain, agitate 1 min, replace with fresh stain, agitate 15 min

Table 5. cont.

4.	Remove gel and place in a new container of distilled H_2O and wash for two min
5.	Place in a new container with 200 ml of fresh developing solution until a background appears
6.	Place gel for 1 h in distilled water. Destain if required.
Wash 1 h in distilled water.	

Table 6. Diammine silver staining of ultrathin-layer gels (from Allen) (23)

1. **Fixation**	20 per cent TCA for 10 min
2. **Reagents**	Glutaraldehyde–Fresh Daily
	0.45 g Sodium Cacodylate
	4.0 ml 25 per cent Glutaraldehyde (Electronmicroscope grade)
	45 ml Double distilled water
3. **Diammine silver**	0.5 ml of 0.5 per cent $Cu(NO_3)2$
	11 ml 0.175 M NaOH (6.9 g/Liter)
	5.0 ml 95 per cent ethanol
	1.25 g $AgNO_3$ in 25 ml double distilled water
	3.5 ml of fresh 25 per cent NH_4OH (Add reagents in order given)
4. **Reducing agent**	694 ml double glass distilled water
	6.0 ml of 1.0 per cent Citric acid
	1.0 ml of 37 Per cent Formalin
	100 ml 95 per cent Ethanol
4. **Stop bath**	10 per cent Acetic acid

Table 6. cont.

STAINING

1. Wash fixed gel for 3–5 min at 50 °C with agitation.
2. Place gel into glutaraldehyde solution in a closed container and agitate for 1.5 min at 50 °C
3. Wash gel in 10 percent ethanol 4 times 3 min for 125μ gels and 4 times 4 min for 250μ thick gels at 50 °C with agitation
4. Place in Diammine Silver in covered container at 50 °C and agitate for 5 min
5. Wash 30 sec. in 10 per cent ethanol
6. place gel into reducing solution at room temperature and observe for color development. If bands do not appear in 1 min, wash the gel in 10 per cent ethanol and recycle through diammine silver for an additional two minutes. Bands, however, should normally appear immediately and their development is easier to control if a white enameled photographic tray is used for the reduction step.
7. The reduction reaction is stopped by placing the gel into 10 per cent acetic acid for 1 min–longer times can cause band fading. The gel should then be placed into distilled water, rinsed and the surface wiped lightly with a cotton pledget to remove any non-visible silver deposits which will darken the gel on storage. Gels may the be dried on a slide warmer and stored. If stored under reduced light the bands remain visibly stable for at least three years.

Table 7. Non-diammine oxidative silver staining (Merril *et al.*) (34).

Stock solutions

1. **Fixation**	20 min in 12 per cent acetic acid– 50 per cent v/v methanol or in 20 per cent TCA
2. **Oxidative Reagent**	0.0034 **M** potassium dichromate and 0.0032 N nitric acid
3. **Ionic silver**	0.012 **M** Silver nitrate

Table 7. cont.

4. **Reducing solution**	0.28 **M** Sodium carbonate 0.5 ml of a 37 per cent solution of formalin per liter
5. **Destaining solution A**	37.5 g NaCl and 37 g of $CuSO_4$ in 850 ml deionized water. Concentrated ammonium hydroxide is added until the precipitate first formed is completely dissolved and the solution is made up to 1 liter.
6. **Destaining solution B**	436 g sodium thiosulfate diluted to 1 liter water

Solutions A and B are mixed 1:1 and then diluted 1:10 for use.

STAINING

1. Following fixation the gels are washed 3 times for 10 min each in a solutin of 10 per cent ethanol and 5 per cent acetic acid (v/v).
2. Gels are placed in the oxidizing solution for five min.
3. The gel is then placed in the silver nitrate solution for a period of 20 min
4. The gel is then placed in the reducing solution and the solution changed at least twice to prevent the build up of precipitated silver on the surface. Development is usually stopped when a light yellowish background appears by placing the gel in a solution of 10 per cent acetic acid. If the bands or spots are not to the desired intensity the gel can be recycled through the silver nitrate.

Table 8. Colored silver stain (from Adams *et al.*) (27)

1. **Fixation**	50 per cent ethanol 10 per cent acetic acid
2. **Silver reagent**	1.9 g $AgNO_3$ per liter
3. **Reducing solution**	7.5 ml 37 per cent formalin in 0.75 N NaOH
3. **Color enhancer**	7.5 g/liter Na_2CO_3

STAINING

1.	Fix gel for 2 or more hours in fixative.
2.	Wash in fresh fixative for 2 hours and again in a 25 per cent ethanol 10 per cent acetic acid solution 2 times for one hour each, then wash 2 times for 1 hour each in a 10 per cent ethanol 0.5 per cent acetic acid solution.
3.	Place gel in silver nitrate solution for 2 hours or more at room temperature.
4.	Wash the gel in distilled water for 10 to 20 seconds.
5.	Place the gel in reducing solution for 10 min and develop to a light yellow background.
6.	Place the gel into the color enhancing solution for 1 hour.

This procedure will also work on ultrathin-layer gels, however, the reducing step time must be shortened to 10 to 15 seconds or the gels will come loose from the backing. Other steps should be reduced accordingly, similar to the times for diammine silver.

A number of the published silver stains are shown in Table 9 along with their general characteristics. All of these stains have been shown to stain respectively proteins, or nucleic acids, or lipoproteins effectively.

Table 9. Comparison of published silver stains

Reference	Oxidation	Aldehyde	Diammine	Sensitivity
Allen (23)	–	+	+	0.1 ng/mm^2
Ansorge (24)	+	–	–	0.2 ng/mm^2
Goldman and Merril (26)	+	–	–	DNA 0.03 ng/ mm^2
Guevera *et al*. (22)	–	–	+	0.03 ng
Guillemette and Lewis (25)	–Cetyltrimethyl ammonium bromide		–	DNA, RNA
Marshall and Latner (19)	–	+	+	0.03 – 1.0 ng/ mm^2
Merril *et al.* (17)	+	–	–	0.02 ng/mm^2
Oakley *et al.* (33)	–	+	+	1 – 10 ng
Poehling and Neuhoff (31)	–	+	+	0.1 – 0.2 ng
Sammons *et al.* (35)	–	–	–	0.1 ng/mm^2
Switzer *et al* (20)	–	+	+	1 ng
Tsai and Frausch (32)	+	–	+	Lipopolysaccharides
Willoughby and Lambert (29)	–	+	+tungstosilisic acid	Protein on Agarose 3–6 ng/mm^2

Silver stain procedures for biological fluids and tissue extracts separated by electrophoretic methods have provided a powerful new tool with which to visualize these materials at the picogram level. They offer an important advantage in that they are in the same order of sensitivity as autoradiographic and radiofluorographic procedures. However, unlike the latter two, they do not reflect protein synthesis effects as do isotope pulse labeling techniques. Similarly in those instances where autoradiography is not practical, such as in the study plasma proteins or in the less concentrated proteins of sweat and urine, the silver stain fills an important previous gap in technology.

These procedures in conjunction with enhanced enzyme-antibody procedures and autoradiography, all discussed later in this chapter, now offer a much broader approach to the study of a variety of biological processes at the molecular level. The phenomena of seemingly specific polychromatic silver staining and the ability to manipulate the color of some proteins by simply altering fixation offers an additional parameter with which to determine new physical-chemical characteristics at the molecular level.

Obviously many of the present observations are empirical and as yet the mechanisms involved are not defined, or at best are ill-defined. However, these objections will undoubtedly; as with any new procedure, slowly be resolved as new information is added in the future. This is particularly true of the present, and often lively controversy, of the quantitative accuracy of the technique; reminiscent of the Amido Black - Coomassie Blue controversy a number of years ago. At present, the practical application of these techniques with their greatly increased sensitivity, in the authors' opinion, far outweigh any objections based on the theoretical shortcomings or present lack of a fundamental understanding of the procedures themselves.

5.5. Enzyme Visualization

An essential advantage of gel electrophresis, particularly isoelectric focusing on thin-layer or ultrathin-layer gels, is that the enzymes remain essentially in a restrictive matrix where they may be reacted with an appropriate substrate and dye *in situ* with minimal diffusion of the enzyme. Reaction rates and diffusion time of the substrate are of course dependent on the thickness of the gel. As mentioned previously, the thinner the gel the more rapid is the pH equilibration and initiation of the enzyme reaction and color development, thus, decreasing diffusion time of the protein zone with an increase in resolution.Again, a compromise situation is sought to give optimal conditions of gel thickness and resolution, since gels less than 250 μ in thickness may present surface problems which may give smearing affects with some systems (36).

5.5.1. Hydrolases

Hydrolases are enzymes which split their substate hydrolytically in the presence of water. For their demonstration very simple color reactions have been developed. Following binding of the enzyme and appropriate substrate, the hydrolysis of the substrate can produce a fluorescence, or the hydrolyzed substrate may be bound in a second step to an azo-dye. The usual methods are with indigo, phenolpthalein, umbelliferone (37–40) and by azo-coupling methods with diazonium salts such as Fast Red TR or Fast Blue RR (39). The indigo-methods utilize an ester of indoxyl which is hydrolyzed by various esterases such as acetylcholine – hydrolase (EC 3.1.1.7) and acetyl choline acylhydrolase (EC 3.1.1.8), examples of which follow:

Indoxyl acetate + H_2O <———> acetic acid + indoxyl
(non-fluorescent) (fluorescent)

Indoxyl + oxygen +Cu^{++} ————> H_2O + Cu^{++} + indigo
(colorless) (blue)

In the phenolpthalene method (41,42), phenolpthalene diphosphate is a frequently used substrate for the acid phosphatases. Following the enzymatic release of phosphoric acid and alkalization red phenolpthalene is measured according to the following reaction:

Phenolpth. + H_2O <————> phenolpth. + orthophosphoric acid

In the umbelliferone method, the ester or glycoside of 4-methyl-umbelliferone produces, after hydrolysis, the strongly fluorescing compound 4-methylumbelliferone. The reaction is very sensitive and is especially useful to demonstrate hydrolases. However, this compound only fluoresces at alkaline pH and many hydrolases have their pH optimum in the acid region. In such cases, one carries out the enzyme reaction at an acid pH and then places the gel in an alkaline buffer to develop the fluorescence (42,43). The reaction of the β-galactosidase (β-d-galactoside-galactohydrolase), (EC 3.2.1.23) is as follows: (44)

4-Methylumbelliferyl-ß-d-glucopyranoside <——> 4- MeUmb + d-galactose

One distinct disadvantage of the umbelliferone method is that the substance is soluble and, therefore, readily diffuses out of the gel. Also, it is not stable in alkaline solution which can lead to background problems.It is recommended that the enzyme bands be assessed as soon after staining as possible. The following enzymes can be determined with the umbelliferone method: Esterases (carboxyl esterase, ali esterase, ß-esterase), aryl-sulfatase, acid-α-glucosidase, ß-galactosidase, α-galactosidase,α-mannosidase , ß-N- acetylglucosaminidase, ß-glucuronidase,α- fucosidase (45) and acid erythrocyte phosphatase (EC. 3.1.3.1) (44) The hydrolases may also be demonstated using azo – coupling methods using naphthyl esters as substrates. After the enzymatic cleavage the free naphthol is azo -coupled to a diazonium salt such as Fast Blue RR or Fast Red TR (43-46). The reaction of leucine – amino-peptidase (EC 3.4.13.9) is as follows:

1. L-leucine-ß-naphthylamid + H_2 <——> ß-naphthol + leucinamide

2. ß-naphthol + diazonium salt ——> diazo-chromophore

The first part of the reaction is enzymatically catalized, while the second part of the reaction is spontaneous. naphthyl esters may be used for the demonstration acetylcholine esterase (48), alkaline phosphatase (49,50), dipeptidyl- amino peptidase (51), and the non-specific acid esterases (47).

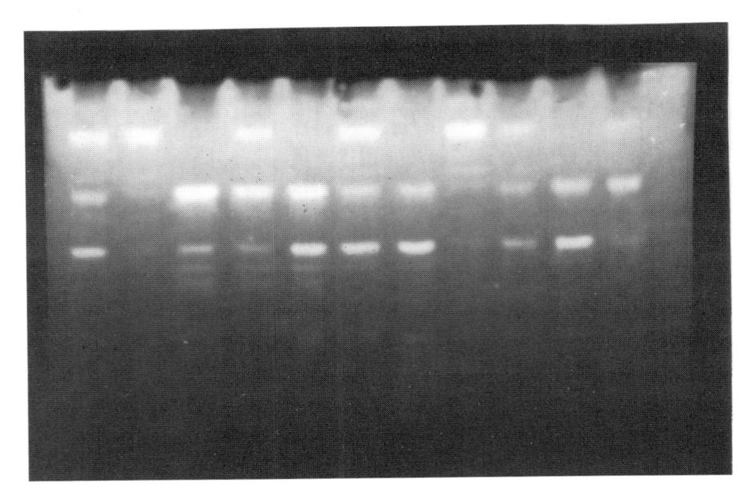

Fig. 3. Erythrocyte acid phosphatases separated by ultrathin-layer isoelectric focusing nd demonstrated by 4MeUmb. Phenotypes BA, A, C, CA, CB, BA, B, A, BA, B and CA; the separation distance was 5.4 cm. (Courtesy Dr. B. Budowle)

5.5.2 Procedures

Table 10. Non – specific esterase method

1. **Stock buffer**	1.0M Tris – Chloride (60.5 g TRIZMA base, 39 ml Conc. HCl)
2. **Reagents**	A. Buffer Tris-Cl 0.2 M pH 6.6 – 96 ml
	B. Substrate alpha-naphthyl-butyrate (1 per cent in acetone) – 4ml
	C. Diazonium salt Fast Red TR – 100mg
3. **Reaction**	Incubate at 37 °C 15-30 min. Stop in 10 per cent acetic acid

Examples of this method on both thin and ultra-thin layer isoelectric focusing is shown in Figs. 4.

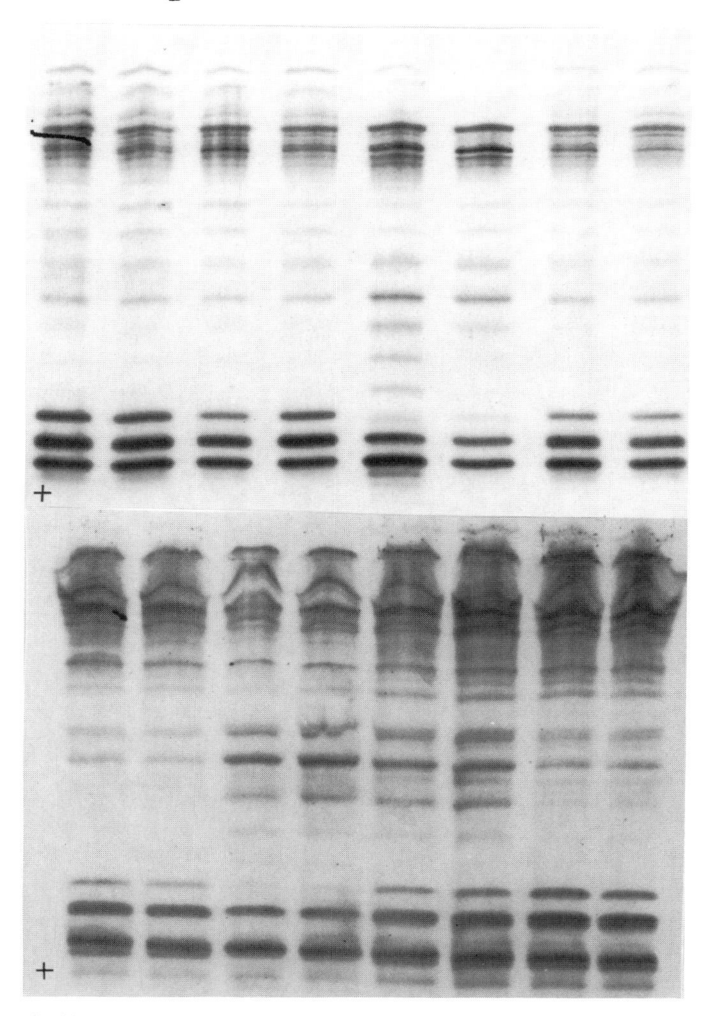

Fig. 4. Mouse kidney non-specific esterases separated by PAGIF on a pH 3 to 8 Ampholine gradient. The substrate was α-naphthyl butyrate and the diazonium salt was Fast Red TR. (A) 1-2, normal male; 3-4, normal female; 5-6, cisplatin treated males; 7-8, diethyldithiocarbamate treated + cisplatin treated males run on an 0.75 mm gel at a maximum if 160 V/cm in 65 min (B) Samples 1-8 in reverse order run on ultrathin-layer in 40 min at a maximum of 375 V/cm.(from Allen *et al.*) (51).

5.5.2.1. Dipeptidyl aminopeptidases

Dipeptidyl aminopeptidases can be demonstrated in a similar manner to the non-specific acid esterases as shown by Allen *et al.* (52) by simply substituting the buffer and substrate. For the demonstration of the multiple molecular forms of this enzyme, which may be demonstrated in soluble macrophage extracts. One merely substitutes lys-ala-4-methoxy-β- naphthylamide as the substrate in 0.005 **M** sodium cacodylate at pH 5.5. The same diazonium salt Fast Red TR may be used as the coupling agent. An example is shown in Fig. 4.

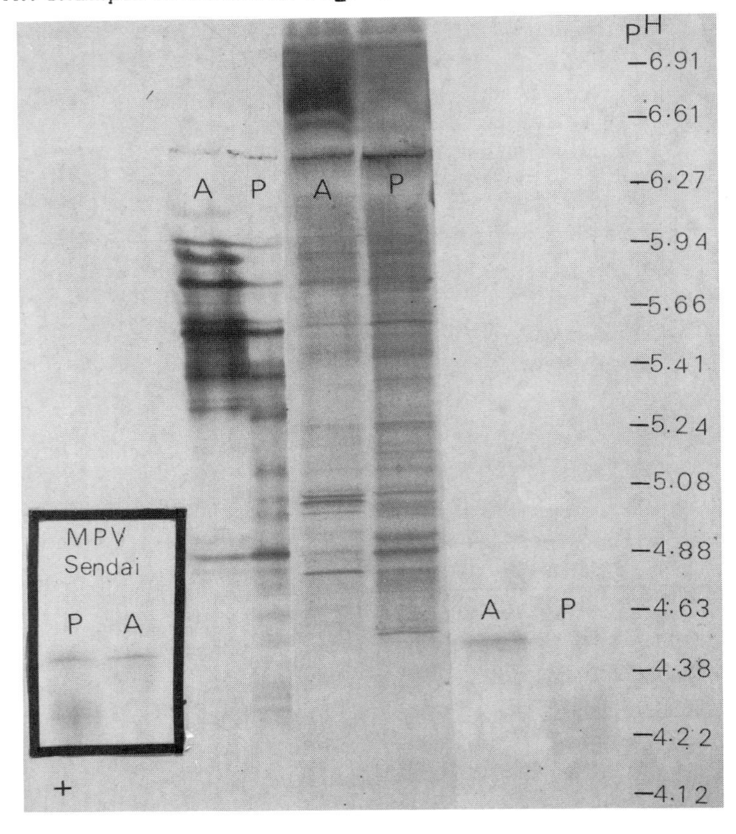

Fig. 5. Comparison of rat macrophage extracts stained for protein, non-specific esterase and dipeptidyl peptidase. Run conditions similar to Fig. 4. Panel (A). Insert shows pattern in Sendai virus infection.(from Allen *et al.*) (52)

5.5.3. Oxireductases

The oxireductases (45) are enzymes which produce reversible equilibrium oxidation-reduction reactions in the general following manner:

$$\text{Substrate.}H_2 + \text{cosubstrate} \longleftrightarrow \text{substrate} + \text{cosubstrate.}H_2$$

Enzymes of this group which use NAD(P) as substrate are usually called dehydrogenases. Such enzymes, using oxygen as an electron acceptor, may be designated also as oxidases.

5.5.3.1. Demonstration of oxidases

Peroxidases (45) catalyze the oxidation of phenols and aromatic amino acids in the presence of hydrogen peroxide. As substrate for these enzymes O-dianisidine is often used. However, this is a carcinogen and thus 3-amino-9-ethyl-carbazol is better to use. (This compound should also be handled with care!) Using the carbazol to detect the peroxidases the following general scheme is used:

3-amino-9-ethyl-carbazol + H^+ + H_2O_2 ⟶ $2H_2O$ + 3-amino-9-ethyl-carbazol.Ox

(Red-Gold) (Brown)

The carbazol can be replaced by pyrogallol which in the oxidized state gives rise to red purpurogallin. Enzyme reactions, during whose course hydrogen peroxide is produced, can be visualized by peroxidase and 3- amino-9-ethyl-carbazol. To this group belong d-aminoacid-oxidase as well as the peptidases A, B, C, E and S.

5.5.4. Dehydrogenases

The dehydrogenases (45) may all be recognized by the general reduction reaction:

$$\text{Subst.} + \text{NAD(P)H}_2 \;\longleftrightarrow\; \text{Subst.H}_2 + \text{NAD(P)}$$

The non-fluorescing oxidized NAD(P) is visible against the fluorescing background of the reduced cosubstrates. This reaction is often used also to demonstrate coupled reactions. For example,to identify triosephosphate-isomerase with alpha-glycerolphosphate-dehydrogenase as a helper system, or the demonstration of 3',5' -cyclic AMP-phosphodiesterase (53). A disadvantage is the excellent solubility of NAD(P), so that the oxidation reaction is favored. The reduction reaction must always then be used if the equilibrium of the reaction is completely driven to the side of the oxidized coenzyme, also the oxidation reaction is not useable. To suitably visualize the oxidation reaction of the dehydrogenases:

$$\text{Substrate.H}_2 + \text{NAD(P)} \longleftrightarrow. \text{Substrate} + \text{NAD(P)H}_2$$

In this reaction the so called tetrazolium salts (54) are used. They are in the oxidized form, water soluble and colorless. In the reduced form they are known as formazans, are colored and for all practical purposes insoluable. Most dehydrogenases cannot directly transport hydrogen to a tetrazolium salt. There must be a reaction system interposed to transport the hydrogen from NAD(P) to the tetrazolium salt. *In vivo* the cytochrome system carries out this task (55). On the other hand, *in vitro* the diaphorase system or N- methyl- phenazine-methyl- sulfate (PMS) (56,57) or Meldol Blue (8- dimethyl- amino-2,3-benzophenoxyazine) can take its place,so that the following general reaction can be demonstrated:

A. $\text{Subst.H}_2 + \text{NAD(P)} \longleftrightarrow. \text{Subst.} + \text{NAD(P)H}_2$

B. $\text{Substrate} + \text{NAD(P)H}_2 + \text{PMS (Ox.)} \longleftrightarrow. \text{NAD(P)} + \text{PMS (Red.)};$

C. PMS (Red) + Tetrazolium salt ———> PMS (Ox.) + Formazan
(no color) (color)

One can use both mono- and di- tetrazolium salts. The mono-salts accept two electrons and the di-salts four. Here the dehydrogenase transports two electrons, where the amount of the formazan produced is in direct proportion to the amount of the reduced substrate. In the case of di-tetrazolium salts two substrate molecules must be reduced for the production of one molecule of the acceptor. Thus, with slower reactions an incomplete reduction may occur and the sensitivity be lessened.For example, incompletely reduced Nitro-BT is red, while the fully reduced compound is blue-black. Also, with smaller amounts of dehydrogenases,the more sensitive mono-salts such as MTT are advantageous.NAD(P) dependent dehydrogenases, which can be demonstrated with PMS and a tetrazolium salt such as NBT are: alcohol- glycerophosphate- sorbitol-, lactic-, malic-, isocitate-, glucose-, glyceraldehyde-3 phosphate-, glucose-1-phosphate-, 6-phosphoglucon-ate-, glutamic-, and shikimic-dehydrogenase (58). To demonstrate the enzyme glutathione-reductase (EC1.6.4.2) one must replace PMS with dichlorophenolindophenol (DCIP),which does not accept the electron from $NADPH_2$ but from reduced glutathione by means of MTT (3-(4,5-Dim-etylthiazolol-2) -,5-diphenyltetrazolium chloride). (DCIP can not be reduced by $NAD(P)H_2$. In this case the course of the reaction is as follows:

A. $NADPH_2$ + glutathione (Ox.) <———> NADP + glutathione (reduced);

B. glutathione (reduced) + DCIP (Ox.) <———> glutathione (Ox.) + DCIP (reduced);

C. DCIP (reduced) MTT ———> DCIP (Ox.) + formazan

As examples of the dehydrogenases the practical formulation for both LDH and MDH (59) are given below:

Table 11. Method for LDH and MDH

1. **Buffer**	0.05 M Tris–Glycine pH 8.4 (6.0 g TRIZMA base, 28.8 g Glycine diluted to a final volume of 1 liter).
2. **Reagents**	84.0 ml Stock buffer
3. **Substrate**	6.0 ml of 1.0 M lithium lactate or L (–) malic acid
4. **Hydrogen acceptor**	2.0 ml (10 mg/ml) NAD
5. **Electron donor-acceptor**	4.0 ml PMS (0.1 mg/ml)
	2.0 ml NBT (0.1 mg/ml)
6. **Reaction**	Gels are stained 30 min. at 37 °C in a light-tight container and reaction stopped in 10 per cent acetic acid. Store in dark at 4 °C

5.5.5. Transferases, lyases, isomerases and ligases

Demonstration of transferases, lyases, isomerases and ligases is accomplished in coupled enzyme tests. The tests are so designed that at the end of a chain of reactions a dehydrogenase or peroxidase reaction occurs. Therefore, a colored complex or a fluorescent substance can be produced in order to demonstrate the desired reaction as having taken place. An example of this type of reaction may be illustrated by the positive stain procedure for GPT (EC. 2.6.1.2) (60).

$$\text{Pyruvate} + \text{L-glutamic acid} \xleftrightarrow{\text{GPT}} \text{L-alanine} + \text{alpha-ketoglutaric acid;}$$

$$\text{L-glutamic acid} + \text{APAD} \longrightarrow \text{alpha-ketoglutaric acid} + \text{APAD.H}$$

$$\text{APAD.H} + \text{PMS} + \text{MTT} \longrightarrow \text{APAD} + \text{Formazan}$$

The general method for this class of enzymes is given below, with GPT as an example of a standard method and PGM as a rapid method on ultrathin-layer gels. In the former, gels less than 1.0 mm in thickness are desirable since, diffusion and background from extended reaction times in thicker gels make this a rather difficult test to interpret for genetic classification:

Table 12. Glutamic pyruvic transaminase

1. **Buffer**	40 ml 0.4 M Tris-HCl pH 8.2
2. **Substrates**	0.6 g L-alanine
	70 mg alpha-ketoglutaric acid
3. **Hydrogen acceptor**	15 mg APAD (Sigma # A5251) or 50 mg NAD
4. **Electron acceptor-donor**	4.0 mg PMS
	6.0 mg MTT
5. **Co-factor**	2.0 mg pyridoxal phosphate
6. **Reaction**	The reaction is carried out in a light-tight container at 37 °C with constant agitation for 30 to 50 min. depending on gel thickness.

The transferases that may be demonstrated using this general type of reaction are as follows: ornithine-carbamoyl-transferase (EC 2.1.3.3), purinenucleoside- phosphorylase (EC 2.4.2.1), adeninephosphoribosyl-transferase (EC 2.4.2.7), glutamic-pyruvate-transaminase (EC 2.6.1.2), hypoxanthine- phosphoribosyl-transferase (EC 2.4.2.8), hexokinase (EC 2.7.1.1), 6, phosphofructo-kinase (EC 2.7.1.11), creatin-kinase (EC 2.7.3.2) adenylate-kinase (EC 2.7.4.3), guanylate-kinase (EC 2.7.4.89) nucleoside-triphosphate adenylate-kinase (EC 2.7.4.10) phosphogluco-mutase (EC 2.7.5.1)phosphoglycero-mutase (EC 2.7.5.3) 2,3-diphosphoglycero-mutase (EC 2.7.5.4) UDP-glucose- phosphorylase (EC 2.7.7.9) and galactose-1-phos-phate- uridyl-transferase (EC 2.7.12)

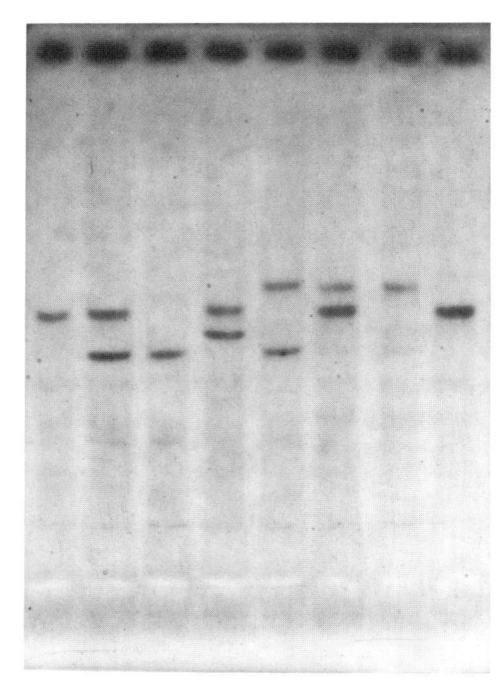

Fig. 6. Phosphoglucomutase separated on a 200 μ gel with an 8 cm separation distance and a maximum of 245 V/cm. From right to left phenotypes 1^+, 1^+2^+, 2^+, 1^+2^-, 1^-2^+, 1^+1^-, 1^-, 1^+. (Courtesy Dr. B. Budowle)

The lyases visualized by this method are: aldolase (EC 4.1.2.13) citrate-synthatase (EC 4.1.3.7), carbonic-anhydrase (EC 4.2.1.1), fumarate-hydratase (EC 4.2.1.2), aconitase (EC 4.2.1.3), enolase (EC 4.2.1.11) and glyoxalase (EC 4.4.1.5). The isomerases are: triosephosphate- isomerase (EC 5.3.1.1), mannosephosphate-isomerase (EC 5.3.1.8) and glucosephosphate- isomerase (EC 5.3.1.9).

Table 13. Phosphoglucomutase: rapid method (from Radola) (61)

1. **Buffer**	From a stock buffer solution consisting of 18 g Tris, 10 g MgCl x $6H_2O$, 1 g Sodium azide per liter; take 1.6 ml and add to 6 ml H_2O. To one half of this solution add 60 mg agarose and dissolve by boiling.
2. **Substrates**	To the other half of the stock buffer add 350 mg glucose-1 -phosphate and 2.5 mg glucose-1, 6-diphosphate
3. **Hydrogen acceptor**	50 mg NADP
4. **Electron acceptor-donor**	10.0 mg PMS 100 µl glucose-6-phosphate dehydrogenase (1 mg/ml) 25.0 mg MTT
5. **Overlay gel**	The boiled agarose is cooled to 60 °C and the substrates and donor-acceptor mixture added and stirred. The mixture is the poured on a GelBond sheet and covered with a sheet of untreated polyester film applied as one would a cover slip to avoid bubbles. The sandwich is then placed in a refrigerator to solidify.
6. **Reaction**	Remove the cover sheet of polyester and lay the gel-substrate-dye agar onto the focused gel, starting at one edge using a rolling motion to assure that no air bubbles are formed and incubate for 2-4 min. at 60 °C. Then remove the agar containing sheet and dry.The enzyme bands appear as blue-violet bands.

Note: The concentration of the reagents in this technique is some five times that used in normal procedures. This allows for the fast reaction time with greatly reduced band diffusion, thus increasing resolution.

5.6. Transfer Methods

There are a number of laboratory procedures that may be employed to perform transfers of a separation to another support medium for immunological, enzyme or other studies. This may be accomplished on a number of supports such as agar, nitrocellulose , cellulose acetate and nylon etc. The printed impression may be reacted with substrate and appropriate coupling agent to demonstrate an enzyme or reacted with specific antibody for an immunolgical probe technique, and in the near future with enzyme labeled antibody probes without the present background problems.

5.6.1. Enzymes

5.6.1.1. Replicate print technique

In the replicate print technique (62,63), the gel following isoelectric focusing is imprinted onto agarose in the following manner: A 0.5mm layer of one per cent agarose is cast on either Gel–Bond (Marine Colloids) or onto a glass plate. The choice here depends on whether the imprinting will be done from an isoelectric focusing gel backed on glass or Mylar, either one flexible and one rigid, or two flexible backed gels are the easiest to handle. Or the gel may be mounted on glass and the agarose on Mylar. To prevent the two gels from sticking together after printing, the agarose gel is first sprayed with a fine distilled water mist. For this purpose a thin layer chromatography atomizer (Desaga) works well. The flexible gel is then carefully pressed onto the focused gel by using a rolling on motion. The resulting sandwich is then pressed firmly together for 1 min. Next, the agarose, GelBond–backed, gel is lifted off and stained specifically for an enzyme. Several prints can be made from the same gel, however, each succeeding print is weaker than the one before and two or three are a practical limit. The original gel may then be stained for protein. This procedure gives the versatility of the older sliced starch gel methods, but with greatly improved resolution.

5.6.2. Membrane impression techniques

Another variation on this theme is the "so called" *Abklatsch* technique described by Radola (64). The dye and substrate are impregnated on a polyamide membrane (Sartorius No. SM 11906) which is reacted directly with the gel containing the separated proteins. This method is carried out at markedly increased reaction temperatures for only one minute with increased substrate and dye concentations for one minute. Due to the short incubation time, little diffusion occurs and the patterns are much sharper.

Table 14. Polyamide membrane procedure (from Radola) (64)

1. **Substrate**	10 mg alpha–naphthyl–acetate in 5 ml acetone
2. **Diazonium salt**	40 mg Fast Blue RR in 15 ml Dist. water

The substrate soaked membrane is applied bubble–free, as in the above method, to the surface of the gel and then incubated for 1 min. at 70 °C in a drying oven. After 1 min, the sandwich is placed briefly in an iodine/calcium iodide solution made up as follows: (12.7 g iodine + 22.5 g calcium iodide in 100 ml of water until both are dissolved and then diluted to 1000 ml). The enzyme bands appear immediately as gold–brown zones on a brown background. The membrane can then be removed, washed in water and then dried with a fan

5.6.3. Nitrocellulose enzyme blot technique

The nitrocellulose blot technique is a modification of the electroblot technique, given later in this section, adapted to thin–layer and ultrathin–layer isoelectric focusing for the demonstration of enzymes. The nitrocellulose membrane is cut to the size of the focusing gel less the area occupied by the electrode wicks. It is then immersed in the buffer with which the enzyme reaction is to be carried out and the rolled onto the glass backed gel in a similar manner to the polyamide membrane technique described previously. Three rectangles of filter paper are then cut to the size of the nitrocellulose rectangle and the first of these immersed in the same buffer as the nitrocellulose.

The second and third pieces are layed on top one at a time and then four to five layers of paper towel are placed on top. Next a 3mm glass plate is layed on top of the resulting sandwich and light pressure applied (500 g). The resulting pressed sandwich is allowed to stand at room temperature for one–three min and then is dissasembled. The nitrocellulose rectangle is reacted in the appropriate substrate, complexing agent and dye. The original gel then may be reprinted a second time for an additional enzyme and the original gel stained for a third enzyme or for protein by one of the methods given under the section on protein staining. It is of interest to note that the printed patterns are always sharper than the original gel. This is also true of the replicate print technique onto agarose described previously. The major difference between the two techniques is that one is on a white background and the other on a clear background. The agarose print technique, therefore, lends itself somewhat more readily to densitometry with standard transmitted light microdensitometers.

5.6.4. Direct immunotechniques

5.6.4.1. Immunoprint

The identification of specific components in complex mixtures of proteins may be obtained by reacting the separated components with specific antibodies following their separation by electrophoresis. This procedure was first described by Alper and Johnson (65) and Ritchie and Smith (66) and in isoelectric focusing by Catsimpoolas (67). These procedures provide the groundwork for a series of valuable techniques. Arnaud *et al.* (68) developed a further and considerably simplified procedure of immunoprinting directly on cellulose acetate membranes to study the α_1–antitrypsin system separated by PAGIF. After completion of electrofocusing the gel a cellulose acetate strip (Sepraphore III, Gelman Instrument Co.) soaked in specific antibody, appropriately diluted in PBS at pH 7.2, is layed directly on the separation track. Care must be taken to assure that no air bubbles are entrapped between the cellulose acetate and the focusing gel. The gel– strip is then incubated at room temperature for one hour. A 3mm glass plate is then placed on top of the gel–strip and a 500 g weight is the placed on top to apply even pressure. After 30 min the weight and plate are removed and the cellulose acetate strip and the gel are washed for 16 hours against three changes of PBS. The strip may be stained directly in Coomassie Brilliant Blue R-250 for 30 seconds (see 5.1.2.1.) and the background destained in destaining solution. To detect the specific immunoprecipitates directly in the gel itself, the gel is fixed in 20 per cent TCA and stained similarly, with the stain time dependent on the gel thickness (one minute in the hot stain is usually sufficient). Precaution must be taken to assure that no high molecular weight proteins remain in the gel. This may be determined by running an adjacent track with the same sample to serve as an unreacted control to assure that no non-reacted proteins remain. The advantage of looking at both the cellulose acetate strip and the gel is that in the case of antibody excess the gel will show a specific immunoprecipitate which might be lost on the cellulose acetate strip. This will also provide a rough titration of the antibody. Any of the many serum proteins, for which specific immunological reagents are available, may be directly analyzed by this technique, an example of which is shown in Fig. 7, where the effect of desialation on the isoelectric points of α_1-antitrypsin allele products is demonstrated.

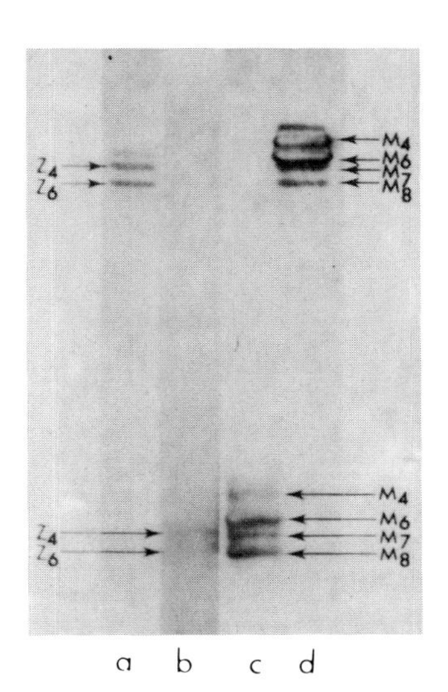

Fig. 7. Immunoprint fixation of MM and ZZ Pi phenotypes. In a and d are shown native Z and M alleles in b and c desialated allele products are shown. (Courtesy Dr. P. Arnaud)

5.6.5 Enhancement techniques

Immunological methods not only provide a high degree of specificity, but also with enhancement techniques can provide a much higer degree of sensitiviry as well. The introduction of the concept of using an enzyme labeled antibody by Nakane (69) and the application of these enzymes to identify tumor associated antigens (70) has led to a rapid development and employment of these techniques in histopathology. Since proteins separated by PAGE, PAGIF and AGIF may be printed or blotted on a nitrocellulose matrix, the application of these techniques to electrophoretic separations is a natural extension. A number of enzymes may be attached to a specific antibody and while the most common one is horseradish peroxidase, glucose oxidase, lactoperoxidase, alkaline phosphatase, lactoperoxidase and β-galactosidase may be employed also as the enzyme. There are four basic techniqes for the application of HRP-labelled antibodies for the conventional immuno-enzyme bridge technique.

5.6.5.1. Direct

The simplest is the direct technique, which uses a horseradish peroxidase (HRP)-labelled monospecific antiserum which binds to the antigenic sites on the gel, or better, following printing or blotting on nitrocellulose. The antibody bound enzyme is then demonstrated by using its specific substrate chromogen reagent diaminobenzidine (DAB).

5.6.5.2. Indirect

The indirect technique uses a similar monospecific antiserum, but in this case, an unlabeled monospecific serum is used. Sites of antibody binding are then identified by the subsequent addition of a second labeled antibody specific for the immunoglobulin of the species used to raise the primary antiserum. ι.ε. horseradish peroxidase (HRP) labeled goat anti-rabbit immunoglobulin serum. The antibody bound enzyme is demonstrated using its specific substrate chromogen reagent , ε.γ.(DAB), analogous to the direct technique above. The indirect technique is five times more sensitive than the direct technique, due to the multiplication of second antibody binding sites (which are provided by the first antibody).

5.6.5.3. Peroxidase anti-peroxidase

Peroxidase anti–peroxidase reagent (PAP) is rabbit anti–peroxidase antibody complexed with its antigen HRP. PAP rabbit immune complex will bind primarily with goat anti–rabbit immunoglobulin previously bound to the antigen, resulting in an increases in sensitivity without the loss of specificity sometimes found associated with multi–step techniques.The PAP technique further increases the sensitivity; however, it requires a third step. As in the indirect method, the antigenic sites bind the monospecific i.e. rabbit antiserum. The PAP complex then reacts with goat anti–rabbit antserum, which is applied next. The goat anti–rabbit immunoglobulin is unlabeled and binds the PAP reagent applied as a third step. The antibody bound peroxidase is then demonstrated with the substrate chromogen DAB.

5.6.5.4. Enzyme bridge

The enzyme bridge technique is similar to above, however, the rabbit anti–peroxidase and its corresponding antigen are applied separately and sequentially, which results in a four step procedure.

5.6.5.5. Avidin biotin method

More recently the development of the biotin–avidin system, which also allows one to study a variety of problems at the molecular level has come into wide use. Thus, lectins which bind with a specific carbohydrate under study, or an antibody to an antigen in very low concentation *i.e.*, in the picogram range can be demonstrated effectively.

5.6.5.6. Biotinylation with Protein A

Protein A isolated from a strain of *Staphylococcus aureus* is a 41,000 molecular weight protein capable of binding the Fc fragment of immunoglobulins.This interaction is not species–specific, as protein A has been shown to bind IgG from a variety of species, but not all. The binding is not restricted to IgG, as studies have demonstrated that

Protein A can interact with IgA, IgM and subclasses of IgG from different species. Biotinylated Protein A is a useful reagent for the detection or localization of immunoglobulin–antigen complexes used with an appropriate avidin–conjugate. Biotinylated Protein A can be employed for the detection of positive clones in monoclonal antibody production using avidin conjugated with peroxidase or other marker enzymes. Thus, biotinylated Protein A can be used potentially in many systems as a substitute for biotinylated second antibody. A schematic presentation of the various approaches is shown in Fig. 8. Due to problems of penetration on gel matices these techniques are best performed following nitrocellulose blot techniques given in the following section. As yet, the background problems with these techniques have not been solved and we therefore, have not given a procedure for their use.

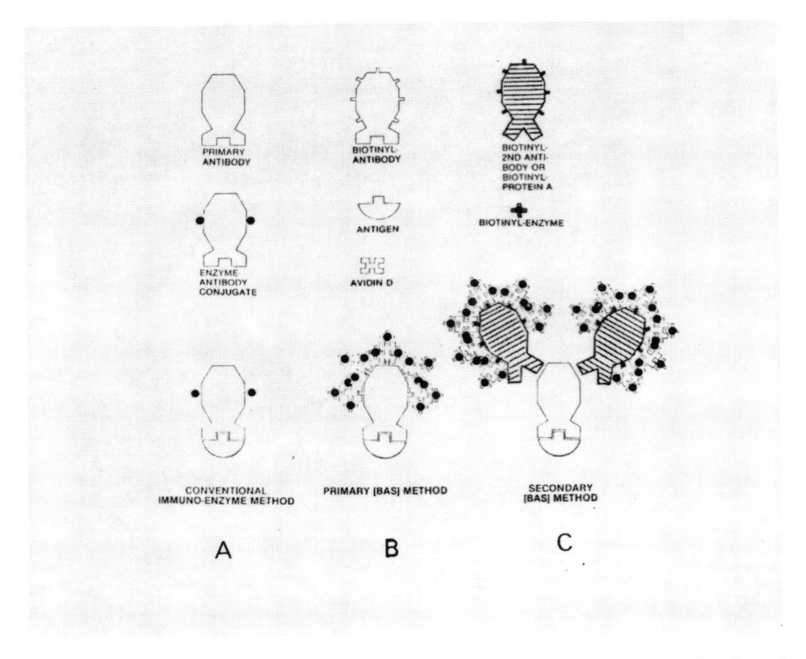

Fig. 8. A schematic presentation of immunoenzyme methods. A, conventional method; B, primary biotin–avidin method; C, secondary biotin–avidin method.

5.7. Electroblotting

Electrophoretic transfer of proteins to membranes or diazotized paper is being used increasingly for the characterization of trace amounts of biologically - important macromolecules. Binding to immobilizing membranes is used when the highest levels of sensitivity, specificity, and reproducibility of immunological reactions are needed.

A wide variety of protein antigens bound to the surface of nitrocellulose membranes have been characterized already. These antigens include subunits of immunoglobulins, albumin subunits, tumor marker antigens, glycoproteins, enzymes, and biologically-active polypeptides. Detection has been either by protein staining, carbohydrate staining, autoradiography, immunoperoxidase staining, or lectin staining. This method offers many advantagers over previous techniques. Several problems which can occur causing confusing results and possible misinterpretation of results are described in this section along with practical solutions.

Immobilizing matrices include nitrocellulose, charged modified nylon and diazotized cellulose papers (DBM and DPT). Nitrocellulose membranes should be employed first in those applications using immunoperoxidase staining. They have the advantages of:

a) It allows protein staining with ponceau S which is a compromise of sensitivity with ease of destaining

b) It has nonspecific binding sites easily blocked with inexpensive and easily applied reagents.

c) It leaves most specific protein antigen sites after membrane binding free for reactions with protein probes.

d) It allows rapid washing from the membrane of unreacted antibodies and other proteins leaving a clean membrane background with increased sensitivity and specificty.

5.7.1. Specific techniques

Peroxidase-conjugated antibodies (see 5.6.5.1.) may be used for detecting picogram amounts protein antigens bound to nitrocellulose membranes. In this technique proteins are first electrophoretically transferred and bound to the nitrocellulose membrane following their separation by agarose or polyacrylamide gel electrophoresis or isoelectric focusing. Following binding of the transferred proteins, the remaining binding sites are blocked with an extraneous protein non-reactive in subsequent steps. The membrane with immobilized proteins is incubated with antibody specific to the antigen determinant (epitope) on the bound proteins which is to be detected. It is then washed removing free first antibody and reacted with a second antibody against the immunoglobulins of the first antibody. The membrane following washing is reacted with enzyme substrate for color development. When chloronaphthol is used, for example, as the substate purple bands are seen against a white background.

The method offers many advantages including presentation of protein antigens on the surface of the membrane for reaction with antibody, thereby, increasing the sensitivity of detection and shortening the test time in comparison with diffusion techniques. In addition, this method of presenting antigens to antibody allows monoclonal antibodies to be used that may not work in single step immunoprecipitation techniques.

Table 15. Nitrocellulose membrane procedure

MATERIALS

1. **Membranes**	Nitrocellulose membranes, Recommended No. BA 83 (0.2µ pore size) Schleicher and Schuell, Keene, NH
2. **Buffers**	Trizma and Trizma hydrochloride Sigma Cat. No. T-1503, T-3253
3. **Second antibody**	Peroxidase-conjugated second antibody (Anti-Rabbit IgG) obtained from Cappel Laboratories (No. 3212-0081), BioRad (provided in kit), and DAKOPATTS (No. P217).
4. **Color**	4-Chloro-1-naphthol, color development substrate, Bio-Rad (provided in kit), and SIGMA (No. C-8890).
5. **Oxidizer**	Hydrogen Peroxide - 30 per cent aqueous solution, stabilized.
6.**Gelatin**	Hipure Liquid Gelatin, (No. 50-005), Norland Products Inc.

Table 15. cont.

7. **Detergents**	Nonidet P–40, nonionic detergent, Sigma (No. N 6507). Tween–20, Sigma (No. P1379), and BioRad (provided in kit)
8. **Alcohol**	Methanol–ACS, Reagent grade.

It is recommended that the nitrocellulose membranes be stored in a cool dark place away from volatile chemicals, that the antibodies be stored at 4 °C, that the color development reagent be stored at –20 °C and that the liquid gelatin be stored at room temperature or in a 30 per cent solution in Tris buffered saline (v/v) at 4 °C.

REAGENTS

1. **Transfer buffer**	Tris–glycine buffer (25mM Tris, 192 mM glycine, pH 8.3). 12.11 g Trizma base 57.68g glycine dilute to 4L with 20 % methanol.
2. **Tris buffered saline**	(TBS) 5.0 g Trizmahydrochloride 0.94 g Trizma base 58.48 g NaCl Dilute to 2 L with high grade deionized water. This yields a solution 20 mM Tris, 500 mM NaCl at a pH of 7.5.
3. **Blocking soln.**	3 % liquid gelatin–TBS 30 ml Hipure liquid gelatin 70 ml TBS Store at 4 °C For use dilute above stock soln. 1 : 10 Add 0.05 ml Nonidet P–40 per 100 ml
4. **First antibody**	Dilute antibody 1 : 100 with TBS containing 1 % liquid gelatin and 0.05 % Nonidet P–40
5. **Washing soln.**	100 ml TBS 0.05 ml Tween 20

Table 15. cont.

6. **Second antibody**	Antibody specific to the species of animal producing the primary antibody is diluted in TBS. Generally the dilution is 1:2000 to 1:4000.
7. **Color developer**	a. 60 mg 4-chloro-1-naphthol in 20 ml methanol b. 60µl of 30 % H_2O_2 in 100 ml TBS

Note: do not use plastic dishes with protein solutions in this technique, as the proteins will bind to the plastic.

Electroblot transfer procedure

1. Prepare quantities of transfer buffer sufficient for transferring proteins from the separation gels to nitrocellulose membranes prior to the actual electrophoretic separation.
2. Fill a tray with the tranfer buffer towards the end of the electrophoretic separation. Submerge one-half of an electroblot support frame, the sponge, two sheets of S & S no. 470 filter paper and one sheet of nitrocellulose cut to the dimensions of the separating gel. The nitrocellulose membrane should be wetted by capillary action, avoiding trapping air in the nitrocellulose pores.
3. Immediately after electrophoresis separation is completed, turn off power and remove the separation gel.
4. Place the separation gel on one of the wet pieces of filter paper and then place the wetted nitrocellulose membrane on top of the gel, excluding any air bubbles. It is important to mark one corner of the gel and nitrocellulose in order to avoid confusion in the subsequent orientation of the gel when it is placed in the electroblot apparatus.
5. Place the other piece of wet filter paper on the nitrocellulose membrane, sandwiching the gel and membrane between the two filter paper sheets.
6. Place the "sandwich" on the sponge in the buffer-filled tray. Place the other half of the electroblot support frame on the "sandwich" and snap the frame together.
7. Place the entire assembly into the electroblot apparatus so that the current flow is from the gel to the membrane *i.e.,* the cathode on the gel side and the anode on the membrane side. Slowly pour the Tris-glycine transfer buffer into the apparatus. Then place a magnetic stirring bar into the tank and place the tank on a magnetic stirrer to continuously circulate the buffer during transfer. Insert a cooling coil into the buffer to maintain the temperature at 10 °C during the transfer.

Table 15. cont.

---

8. For most proteins, the recommended voltage gradient is 1.5 V/cm for 30 min and then 4 V/cm for 1–3 h.
9. Disassemble the support frame at the end of the electrophoretic transfer and remove the nitrocellulose membrane.

Detection step

1. The nitrocellulose membrane with the transferred proteins is placed on a piece of hard filter paper, such as S & S no. 577 and air dried for 30 min or more at room temperature.
2. The membrane is picked up by forceps and briefly allowed to touch the blocking solution with the edge of the membrane so as to draw the solution into the membrane by capillary action. When fully wetted the membrane is placed into the blocking solution and gently agitated at room temperature for 1 h.
3. The membrane is removed from the blocking solution immediatly after soaking and placed into the first antibody solution for 1 h at room temperature with gentle agitation, or for 16 h at 4 °C. (Note: low affinity antibodies work best at 4 °C.)
4. Remove the membrane from the primary antibody solution, briefly rinse with deionized water and place in the washing solution for 30 min with gentle agitation, then into a fresh second wash solution for an additional 30 min.
5. The membrane is placed in the second antibody solution and incubated for 1 h with gentle agitation.
6. Briefly rinse the membrane with deionized water and wash for 30 min twice with agitation as in 4 above.
7. Place the membrane into freshly prepared color developermix a and b, in which color should appear within 15 min. Nonspecific staining will develop with time and has to be accounted for by the inclusion of proper controls (see hints section below).
8. Color development is stopped by immersion of the membrane into deionized water for 10 min.
9. Air dry the nitrocellulose membrane between hard filter paper sheets, (*e.g.* S & S no. 577).
10. The membrane can be photographed either wet or dry and can be made translucent by saturation with immersion oil, or temporarily cleared while wet, with tetrachlorethylene.

5.7.2. Operating Hints

Electrical current can not pass through the commonly used supports (usually a treated Mylar film) and the separation gel when they are placed unsupported in the electroblot apparatus. Electroblot transfers of protein separation patterns are accomplished readily when the gel matrix is cast on a porous support (70). A porous, high density, hydrophilic, polyethylene support makes handling flexible gels easier and minimizes gel distortion. This allows assembly of the transfer sandwich in air, where previously it was necessary to assemble it under the buffer surface. A transfer buffer-saturated sponge is placed on one-half of the plastic grid transport support, and in order, a tranfer buffer-wet thick filter paper placed next on the sponge, followed by the sintered support and gel. A buffer-wet nitrocellulose membrane is the rolled onto the gel surface and the second piece of buffer-wet filter paper layed on, followed by the other half of the grid transport support. The assembly is placed directly in the electroblotting apparatus, containing transfer buffer. Note: the transferred separation patterns are superior to those obtained when assembly of the sandwich is performed under the buffer surface.

Following transfer of the proteins to the nitrocellulose membrane, the membrane is air dried. Air drying at room temperature overnight significantly improves the detection of tumor markers, particularly those of low molecular weight, such as Interleukin I.

Electroblotting at too high a field strength frequently causes loss of protein binding to nitrocellulose membranes. If examination of both sides of the membrane reveals more of the low molecular weight proteins binding to the far side of the membrane, this indicates that delayed binding is occurring. Accordingly, lower field strengths are required initially to bind these proteins. Normally this problem may be eliminated by using 6 V/cm or less. Nitrocellulose membranes with a pore size Of 0.2 μ rather than 0.45 μ will improve binding of low molecular weight proteins. In any event, protein binding will differ on both sides of the gel to some degree. Thus, clearing of the membrane is required to determine this quantitatively. Wetting the membrane with water or immersion oil permits photography, but leaves much to be desired from a quantitative standpoint. Tetrachloroethylene (suggested by Schleicher and Schuell) results in a clear membrane until it evaporates.

It must be handled with care, since it is harmful if inhaled.

5.7.2.1. Causes of protein loss

It has been observed that protein initially bound to the membrane sometimes is lost in subsequent processing steps, thus, the binding of proteins to nitrocellulose cannot be assumed to be totally complete or irreversible. Hanging the membrane in a sealed chamber saturated with glutaraldehyde vapor for 0.5 to 24 h increases the amount bound and the detection limit of carcinoembryonic antigen (70) as well as subunits of human albumin. Glutaraldehyde treatment should always be followed by sodium borohydride reduction of any residual aldehyde groups. It has been found that the extended air drying, described above, is also more efficacious, and simpler, than glutaraldehyde treatment.

Several proteins examined, notably iron-containing ones, such as hemoglobin and haptoglobin from hemolyzed serum, have the ability to react with the substrate/chromogen in a manner analogous to the peroxidase labeling technique. This nonspecific reaction can be eliminated by sequential treatment with 3 per cent hydrogen peroxide (10 min), 2.8 per cent periodic acid (5min) or 0.02 per cent sodium borohydride (2 min) prepared in Tris-buffered saline. This treatment is performed following the initial transfer of protein onto the membrane.

Nitrocellulose membranes only are utilized in the procedures given in this section. This is due to the fact that most investigators working with membrane binding find that the nitrocellulose membranes have the most desired properties for binding proteins to be used in subsequent immunological reactions. Other membranes (nylon, charged paper, *etc.*) frequently bind proteins so tightly that it is difficult to clear the background of unreacted reagents. Perhaps the use of blocking agents, such as liquid gelatin (described below) will allow more efficient use of such supports for increased binding of trace proteins.

The question naturally arises why tranfer the separated proteins when the dried (collapsed) gel matrix, on which the proteins were initially separated, can also be reacted with antibody probes (71). Protein antigens in dried gel matrices were found to be relatively inaccessible to antibody in comparison to nitrocellulose and Arnaud (72) has found, using isotopic labeling, that tranfer from ultrathin-layer PAGIF is more effective by pressure blotting than by electroblotting.

5.7.2.2. Background staining

One of the biggest problems associated with using nitrocellulose or othe membranes for protein transfers and immunoprobes is nonspecific binding of the antibody probesthemselves to the membrane. A small amount of residual nonspecific staining can be of significant importance when working with low levels of sample protein that need to be detected.The sensitivity of detection systems has improved so markedly, that small amounts of nonspecific binding, either not blocked by the protein additives or contributed by the additives themselves, become major obstacles in the development of new assay systems. A great amount of effort can be expended to overcome this problem, often without success. Procedures for blocking nonspecific binding sites on the nitrocellulose membrane include the addition of normal serum protein, bovine serum albumin and gelatin. Normal serum protein normally represents proteins isolated from plasma by a variety of isolation procedures and shows significant variation in individual protein quality and quantity, as well as in stability of biologically labile proteins. Bovine serum albumin, mainly isolated using cold ethanol (Cohn Fraction V), contains other proteins in addition to albumin. There are significant variations in the albumin, the stabilizers added in processing and the quality of the albumin from different manufacturers. Animal gelatin currently used in commercial kits for immunological detection *via* binding to nitrocellulose membranes (*e.g.* ImmunoBlot, BioRad) has to be warmed to 37 °C to dissolve and for the gelatin solution to remain fluid. Since animal gelatin gels at 4 °C, antigen-antibody reactions containing gelatin in the buffer cannnot be incubated at 4 °C, although, this is desirable to increase total antibody deposition and to minimize bacterial growth. This is especially true when the reactants require many hours of incubation. Hipure Liquid Gelatin from Norland Products does not have to be heated to dissolve it. It also may be refrigerated in a 30 per cent stock solution (v/v in Tris-buffered saline) without gelling, it can be used in an incubation buffer a 4 °C and it has superior blocking qualities in eliminating background and nonspecific immunological staining. This material is extracted from fish skin, which has a far different proportion of the basic amino acids as compared to animal gelatin. The gelatin comes with the preservative methyl/ propylparaoxybenzoate, which does not interfere with peroxidase staining when used as a one per cent solution. Using this material, background staining of the nitrocellulose membrane may be reduced to insignificant levels, resulting in an increase in the specificity of immunoassays, including Southern (74) blot and dot applications (71-73).

5.7.2.3. Antibody associated problems

Although liquid gelatin significantly decreases nonspecific antibody binding to the membrane, the antibodies themselves may cause problems. It is germane to mention that many commercial enzyme conjugated antibodies are notoriously nonspecific in action; showing significant cross reactions with other antigens as well as binding nonspecifically to the membrane bound proteins. The practitioners of this highly sensitive technique will have to screen the "so called" *affinity purified* reagents before their use in the newer immunoassays. Although background subtraction techniques are not desirable, they are frequently required for most of the recently developed immunoassays when existing commercial antibody probes are used. As a consequence, investigators can either peroxidase label their own antibodies, or contact the manufacturer and convey their specialized needs directly. Labeled Protein A can be used, but it is less satisfactory than second antibody, since Protein A does not recognize IgG of all species or IgG subtypes. In addition, its binding is not polyvalent as is that of second antibody (see section 5.6.5.6.).

5.7.2.4. Reagent testing

If no color is produced when substrate/chromogen solution is added to the washed conjugate (after the addition of the peroxidase labeled second antibody and incubation for 25 min), the following procedure may be carried out to test the reagents: 1) Mix 1 ml of subtrate/ chromogen solution in a small glass tube with 1 ml of peroxidase labeled, diluted antibody to determine if there is a problem with either the chromogen solution or the labeled antibody. If there is a color change in the tube, it indicates that no labeled antibody bound to the antigen on the membrane. 2) Mix 1 ml of the chromogen solution with a drop of dissolved peroxidase; if there is color development it indicates that the substrate/chromogen solution is good and that there probably is a problem with the peroxidase label of the second antibody. It should be noted that 4-chloro-1-naphthol is as sensitive as diaminobenzidine, produces less background and is not a proven carcinogen.

5.7.2.5. Troubleshooting

It is recommended that each of the individual procedures manuals from the binding membrane manufacturers be read in detail prior to using the product. The comments above have been included to pass along some of the hands on experience of one of us (C.A.S.) that we feel will simplify the use of these procedures and help to resolve some of the problems one may encounter. It must also be emphasized that this field is rapidly expanding and, therefore, much of the above could be outdated even as this book is in press.

Table 16. Troubleshooting hints for membrane electoblot techniques

Problem	Cause	Solution
1. No reaction or weak color	a. Enzyme inactivation by NaN_3 in buffer	a. No azide, store fresh solutions at 4 °C
	b. Development reagent poor	a. Make immediately before use b. Store at -20 °C in the dark
	c. Antibody not HRP labeled	a. Label in-house
	d. Weak first antibody	a. In-house production
	e. Poor antigen binding	a. Glutaraldehyde fixation b. Air dry 24 h c. Lower voltage d. Better disclosing agents
2. Background	a. Bad blocking reagent	a. Use liquid gelatin
	b. Precipitate in antibody	a. Centrifuge antibody

5.8. Autoradiography

5.8.1. Direct

The use of radioisotopes to label specific proteins was, before the development of the silver stain methods, the most sensitive detection method for the visualization of protein available. However pulse labeling of cells may produce different patterns due to the fact that one is measuring, protein synthesis rather than the normal homeostatic state of the cell and patterns may differ from those stained with the equally sensitive silver stains (34). In some cases autoradiography is as sensitive, or slightly more sensitve than silver *e.g.* with [^{35}S], while [^{125}I], [^{14}C] and [^{32}P] several magnitudes less sensitive (75).

When it is desirable to label cell samples, [^{35}S], [^{14}C], or [^{32}P] are preferred to [^{3}H] because of their greater efficiency. The procedure is the same as for unlabeled samples. Gels are first fixed and stained with Coomassie Brilliant Blue R250, destained and photographed if required. The gels, either 1 or 2-D are then rinsed well through two changes of distilled water with agitation to remove any residual acetic acid. The gels are then softened by soaking them in 2 % glycerol for one h, with at least two changes. The gels are then dried onto filter paper in a vacuum dryer (*e.g.* Bio-Rad model 224). The dried gels are then processed for autoradiography as directed in table 17.

Table 17. Autoradiography procedure (76)

1.	Label dried gel with the appropriate identification in radioactive ink ([^{14}C] in upper right hand corner.
2.	The stained dried gel is placed in direct contact with the X-ray film clamped in a radiographic cassette or between glass plates using metal clips. It is then wrapped in black plastic or placed in a light tight box. After exposure the film is developed according to the manufacturers recommendations.

Direct autoradiography of dried gels with Kodak XAR X-ray film will provide a film image absorbance of 0.02 A^{540} units (a level just visible above background) in a 24 h exposure with 6,000 dpm/cm2 of [^{14}C], or [^{35}S], 1,600 dpm/cm^2 of [^{125}I] and 500 dpm/cm^2 of [^{32}P] (77)

5.8.2. Fluorography

This procedure is used when the sample has been labeled with [^{3}H], [^{14}C], or low levels of [^{35}S]. Gels are run as usual, stained with Coomassie Brilliant Blue R250 and then destained and photographed and the developed for fluorography as shown in Table 18.

Table 18. Fluorography technique (76)

1.	Put gel into DMSO I (dimethyl sulfoxide) with 250 ml per gel (5 Gels Maximum) and agitate for 30 min. Return DMSO I into original DMSO I container. After three uses it sould be discarded.
2.	Place gels into a second (DMSO II) solution with again 250 ml/gel and agitate for an additional 30 min. Return DMSO II to its container, after 3 uses it should be labeled DMSO I.
3.	Place gels in a 13 % PPO (2,5-diphenyloxazole) made up in DMSO. Agitate for 3 h. Return PPO solution to its container. This solution may be used 8 times, after which the PPO must be recovered and the remaing DMSO discarded appropriately as organic waste. PPO should not be put down a drain, since it precipitates in water.
4.	Place gels in distilled water (250 ml/gel) and wash with agitation for 15 min, change water and wash for an additional 15 min with agitation.
5.	Place gels in a 2 % glycerol solution and wash with agitation for 15 min. Change to fresh glycerol and was for an additional 15 min with agitation.
6.	Dry gel under vacuum as in Table 17
7.	Put gels on Kodak X-Omat R™ film that has been flashed. Then place gels with films inside black plastic bags, evacuate air, and heat seal. Store sealed bags at -80 °C. Thaw and develop according to manufacturers directions.
1.Flashing	Pre-exposure of the X-Ray film is performed with an electronic flash unit *e.g.* Vivitar 283. The duration of the flash should not exceed 1 msec, since longer flashes will fog the film without hypersensitizing it. A full charge on the capacitor is assured by allowing the ready light to be on for at least 30 seconds before the flashing procedure is carried out.

Table 18. cont.

2. Filtering	An infra red filter is placed nearest the flash unit to protect the the secondary filters. A deep orange Wratten No. 22 filter is placed next to the infrared filter and then both are covered with Whatman No. 1 filter paper to diffuse the lihgt in order to expose the film evenly.
	PPO may be recovered from the DMSO by adding 1 volume of the PPO in DSMO to 3 volumes of 10 % (v/v) ethanol. After 10 min, the suspension is filtered and the PPO precipitate is washed with 20 volumes of water and then air dried. (78)

Chamberlin (79) has introduced the use of sodium salicylate as the fluor instead of PPO. An advantage is that it is considerably less expensive and that it is water soluble so that DMSO is not required. More recently New England Nuclear has introduced EN^{3}HANCE™ and ENLIGTNING™ as fluorography enhancers which do not require the use of DMSO. The formulations of these proprietary materials are not given.

5.8.3. Dual labeling

McConkey (80,81) has described a method for double labeling. The gel is prepared for fluorography and exposed to X-Ray (KodaK XAR™) film which is sensitive to the photons produced and thus detects both [^{14}C] and [^{3}H]. The gel is then placed in contact with Kodak Direct Exposure Film (81), which is insensitive to the light produced by fluorography and detects then only the [^{14}C] by direct autoradiography. Comparison of the two films allows one to identify either or both of the isotopes and the proteins or polypeptides labeled with each.

5.9. Miscellaneous Stains

5.9.1. Crowle's double stain

Table 19. Crowle's double stain for agarose (82)

1. Fixative	180 ml methanol 30 g Trichloacetic acid 90 g Sulfosalycylic acid 300 ml Distilled water
2. Stain	2.5 g Crocein Scarlet 150 mg Coomassie Brilliant Blue R250 50 ml Glacial acetic acid 30 g Trichloroacetic acid Dilute to 1 liter with distilled water
3. Destaining solution	3.0 ml Glacial acetic acid diluted to 1 liter.
4. Staining	Place the dried gel backed on GelBond into the stain solution at room temperature for 10-15 min with gentle agitation.
5. Destain	Rinse the stained gel in distilled water and then place it into the 0.3 % acetic acid destaining solution. Final rinse the gel in distilled water and dry at 60 ºC.

5.9.2. Lipoprotein prestain.

Plasma lipoproteins may be prestained prior to electrophoresis with Sudan 4-B. This procedure provides a rapid method with PAGE to discriminate the various types of familial hyperlipoproteinemias.

Table 20. Lipoprotein pre-stain (83)

1. Stain solution	25 mg Sudan Black B 24.4 ml Ethylene glycol 0.625 ml 1.5 M Tris-citrate buffer (pH 9.0) Heat for one h and the filter through whatman No. 1 filterpaper.
2. Sample staining	To 50 µl of plasma add 50 µl of the buffered stain. Mix well and incubate for 15-30 min at 37 ºC. Centrifuge 1 min in an Eppendorf Microfuge™ and use supernatant 20 µl as sample.

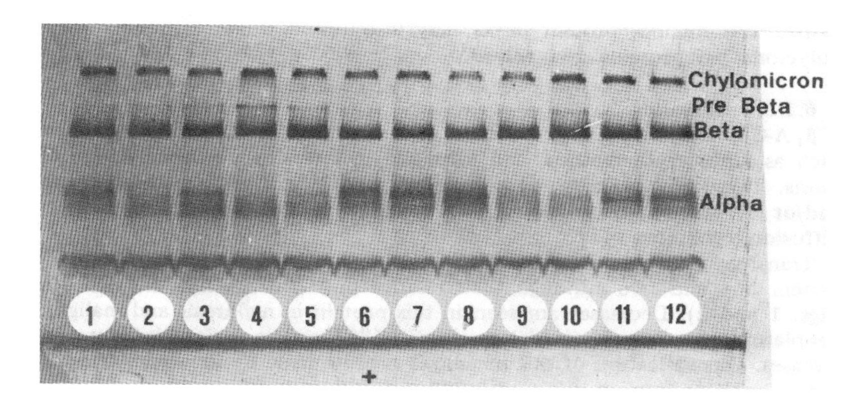

Fig. 9. Separation of pre-stained serum lipoproteins. The separation was carried out in a 3-6-9 % T gradient at a continuous pH of 9.0 in a citate-borate MZE buffer. Sample 1, 2, 3, and 11 show the typical HDL patterns seen with this system. Samples 6-8 are female, marked with the increased HDL and sample 4 shows an increase in the pre-β band. Albumin is marked with Bromphenol Blue tracking dye. (from Allen) (83).

5.9.3. Heme binding proteins

The heme binding proteins such as haptoglobin can be prestained by the addition of hemoglobin to the sample. In order to prepare the hemoglobin prepare a packed cell volume of 5 times washed red blood cells. Lyse the red blood cells in 4.75 ml of distilled water, centrifuge and store supernatant hemoglobin at 4 °C. Add 5 µl of hemoglobin to 40 µl of plasma, which should give approximately 1 µg of hemoglobin per 15 µg plasma protein. Following the separation the heme-binding proteins are demonstrated by disolving 0.05 g of diaminobenzidine and 0.2 g barium peroxide in 5 ml of distilled water. Then add 5.0 ml of glacial acetic acid and pour on the gel surface. The reaction is immediate, with an initial blue-green color which changes to dark brown. Since benzidine is a carcinogen all staining steps should be carried out with rubber gloves and the stain and first water wash of the stained gel discarded appropriate for organic chemical waste.

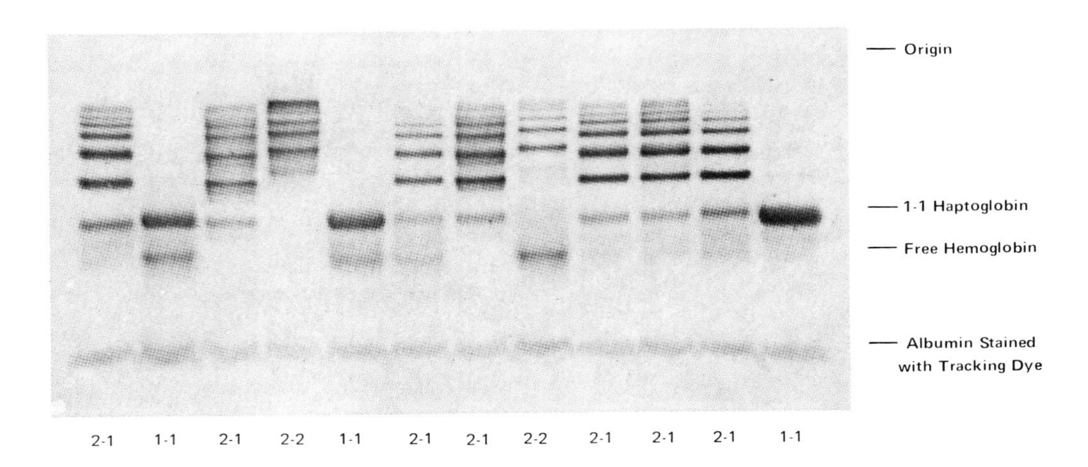

Fig 10. PAGE separation of haptoglobin phenotypes stained for hemoglobin with diaminobenzidine and peroxide. (from Allen) (84)

5.10. Densitometry

One of the most frequently asked questions from a broad spectrum of investigators using PAGE, PAGIF and 2-D procedures is: what is the best microdensitometer for my application and what characteristics are most desirable in a microdensitometer? With the increasing use of two-dimensional techniques the problem is even more magnified. We have attempted to bring together the most important desirable characteristics of such instrumentation in order that the reader may make a choice based on his or her particular needs, with the full understanding that price will probably be an important factor in the choice.

5.10.1. Single dimension scanning

For single dimension scanning a microdensitometer should have the following characteristics: 1) The effective slit width should be no wider than one-quarter of the narrowest band to be scanned. 2) The signal response time should be as small as possible. 3) In instruments with hardwired computers built in, or in those with software and microcomputers, the scan speed should allow at least 11 points to be sampled on each peak. 4) Columnation of the light source and the detector should be preset or adjustable in those instruments using transmitted light. 5) A variable wave length light source is desirable, as is a UV source. 6) The gel transport mechanism should be observed carefully to assure that it is properly designed so that play or looseness will not develop after continued use. 7) The readout device should be able to detect peak location as well as peak density. 8) The instrument should have a dynamic range of at least 3.0 OD units.

Additional features , such as automatic calculation of peak area, peak height, area/height, individual peak area as a per cent of the total area and normalization programs to provide quantification of each peak in mg are also desirable. These all bring into question the performance of the instrument in relation to its cost.

High voltage gradient isoelectric focusing and silver staining can resolve protein bands under 100 microns wide separated by as little as 50 microns. Such resolution will require a light slit or spot of 25 microns, or less, to resolve fully such bands. This may be obtained also by the use of microscope optics in the system to decrease the effective slit width, or by the use of laser scanners. Both increase the cost and the laser types are presently limited to a single wave length of light. Further, small slit widths may decrease the signal intensity to the detector increasing the noise to signal ratio. Detectors are usually photomultiplyer tubes, photo diodes or phototransistors. The first is generally slower and the stability is not the best,. while the photo diode is up to an order of magnitude slower in response time than the photo transistor. In practice, with all the other electronics required the photo diode may only be 2-5 times slower than the phototransistor.

Another question is the choice of a tungsten light source as opposed to a laser. The former allows one to select wave lengths in the visible spectrum, while in the latter, one is limited normally to a single wave length. Also with a tungsten source, one may operate the system under reduced light with the gel open to view for light spot positioning and visualization prior to and during the scan. On the other hand, the laser source instruments require a light tight box, which precludes observation of the gel during the scanning procedure. The light source geometry itself will play a major role in the inhrent noise produced in the system. We have found that a point light source is generally better than an elongated slit and is easier to properly columnate.

The general practice today, of mounting gels on either glass plates or mylar and drying them before scanning, has markedly reduced the problem of lensing effects found in scanning wet gels. Silver staining in conjunction with isoelectric focusing and molecular weight by PAGE and SDS-PAGE have reduced the background and background smear effects of earlier PAGE serum techniques with Coomassie Brillant Blue staining. On the other hand, mylar-backed gels may produce some undesirable background noise not found on gels backed on glass plates.

The rapid proliferation of micro and personal computers in the last two years has opened up an attractive alternative to the dedicated hardwired computer microdensitometers costing over 10,000 dollars. A number of programs are available that can be run on personal computers. With the addition of an analog to digital conversion board, a personal computer such as the Apple II, used to produce this book, can also serve as the processor, readout and computational device for a relatively inexpensive transducer consisting of light source, photo transistor, associated electronics and a transport device for scanning the gel. Thus for under 5,000 dollars one can put together a rather sophisticated micro densitometer, using an inexpensive EC Apparatus Corp. (St. Petersburg, FL 33709) scanner as the transducer. One then has a process controller, word processor and database management system all in one instrument. Obviously this is a personal preference and for those investigators less prone to gadgeteering, a dedicated microdensitometer is more practical. However, we will mention two programs that we have experience with that meet the requirements for single dimension peak evaluation. The first is that of Yakin *et al.* (85) and the second is that of Zeineh and Abdul-Karim (86). The latter system is available commercially (Biomed Instruments Inc., Fullerton, CA 92631).

A number of other densitometers are available commercially for single dimension scanning that employ various automated readout systems. These are: 1) That available from LKB (Gaithersburg, MD 20877), 2) Helena Laboratories (Beaumont, TX 77704), 3) Hirschmann (8025 Unterhaching, FRG) and 4) Desaga (6900 Heidelberg, FRG). The last is based on a Shimadzu design.

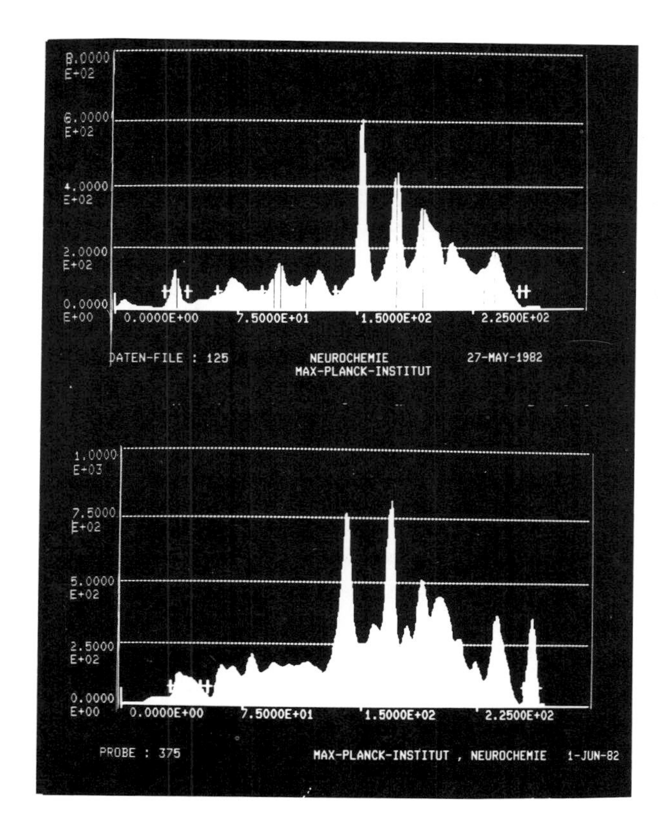

Fig. 11. Densitometric output from a microscope based densitometer and program described by Yakin *et al.* (85). The traces are from the separations shown in Fig. 4. on page 75, panels A and C. (courtesy of Prof. V. Neuhoff)

5.10.2. Two-dimensional densitometry

Two commercial densitometers are presently available for two-dimensional gel scanning. These are about an order of magnitude more expensive than those required for single dimensional scanning. At present there are a two approaches to the transducer and readout systems.These are: 1) The use of of a matrix diode array and 2) the use of a monochromatic laser. The first of these, produced by BioImage Corp. (333 Parkland Plaza, Ann Arbor MI 48106), utilizes a solid state photo diode scanning array of 1728 elements with a resolution of 8 bits for 256 gray levels. This type of instrument measures both reflectance and transmittance and translates this information into density through the computer software. This procedure makes it more difficult to obtain accuracy at higher optical density levels. In addition there may be more optical flare in this type of system than in one using a laser source. On the other hand, this instrument has available a color processing option which allows it to handle colored silver stains also.

A second densitometer for 2-D scanning was announced just as this book was going to press. This instrument is produced by Image Analytics Corp. (Box 362, Hockessin DL 19707). This instrument has a laser light source and measures density directly. The instrument uses a 12 bit processor, as opposed to eight, and is thus capable of 4096 gray levels. This capability allows a better analysis of overlapping spots, but does not have a color capability (87). Projected prices of both instruments are somewhat less than 100,000 dollars.

Both instruments use 16/32 bit processors and are capable of storing 25 to 30 gel patterns in memory and to compare any pattern obtained with a standard, in order to detect new spots automatically. This means a memory storage capability of a least 27 megabytes in hard disk drives. Such instruments have the capability of completely analyzing a complex 2-D gel in about five minutes. Helena systems offers an enhanced version of their single dimension instrument which sequentially scans a 2-D gel in one dimension, thus providing a 2-dimensional presentation. The Biomed model is also obtainable with 2-D capability.

Eikonix (Eikonix Corp., Bedford MA 01730) has recently announced the EIKONIXSCAN™ 78/79 series of image analyzers, which purportedly can replace both the vidicon camera type image analyzer, as well as a microdensitometer. This instrument has a resolution potential of 2540 lines per inch on a sample size of 10 microns. An appropriate micro computer is supplied separately by the user.

5.11. Photography

Photographic techniques for electrophoretic separations appear to cause a number of investigators considerable consternation. The question of which film, which camera and lens are most often asked questions. We have used, for the majority of illustrations in this book, two basic systems with good success. As a light source we have employed a standard X-Ray viewing box with a daylight fluorescent bulb and an opaque white plastic top. For black and white photography we use a Mamiya RB 67♂ single lens reflex with a 90 mm lens and a yellow filter with positive negative Polaroid #665 film. This film has a rating of 75 ASA or 20 DIN. The positive resolution is 14-20 lines /mm and the negative has a resolution of 160-180 lines /mm. The 7.3 x 9.5 cm format, however, is not as desirable as a four by five inch or (10.2 x 12.7 cm) format. The lens on this camera is far superior to that normally found on such systems as the Polaroid MP™ series. The use of close up lenses, as required, more than makes up for the less desirable format size. The advantage of the P/N black and white films is that the positive can be viewed immediately, and if satisfactory, the negative will print even better on number 2 or 3 paper. While shooting more than one picture on the larger format may be more expensive, the time saved in obtaining good prints is worth the additional small cost.

The light source decribed above, is adequate for black and white film, but it will give a greenish-blue background with color film due to the color temperature mismatch. The use of a light magenta filter Tiffen FLD or Hoya FL-W compensates the color temperatures and prevents a green background from appearing in color slides. For color photography, we have used a number of single lens reflex cameras with built in light meters with good success. A 50 or 55 mm lens with the above filters and appropriate close up lenses provides good color slides either with 200 daylight Ektachrome film, or with Kodak photomicrography color film #2483. The latter is used at a film speed of ASA 6-8 also with a Tiffen FLD filter or Hoya FL-W. This film has almost no grain and also makes good black and white enlargements, as well as excellent slides.

For those whose budget is limited to one camera, we would suggest a good 35 mm single lens reflex with an uncoupled, built in light meter. For color photography ASA 200 daylight Ektachrome film, so that the f stop may be kept between f 5.6 and f 11. For black and white, a fine grain film, such as panatomic X or Ilford Pan F in conjunction with a high contrast glossy paper, or the color 2483 photomicrogaphy film for both purposes. As a note of caution the slow color film requires a solid vibration-free stand and mount for the camera due to the longer exposure times . With a little practice gel photography becomes a simple procedure from which excellent results readily can be obtained.

5.12. REFERENCES

1. Raymond, S. and Weintraub, L. : *Science, 130*, 711 (1955).
2. Ornstein, L. and Davis, B. : *Disc Electrophoresis Parts I and II*, Distillation Industries, Div. Eastman Kodak, Rochester N.Y. (1962).
3. Busse, V. : *Z. Klin. Chem. Biochem., 6*, 273 (1969).
4. Radola, B. J. : *Electrophoresis, 1* 43 (1980).
5. Saravis, C. A. : *Electrophoresis '83, Boston, Abst.* (1983).
6. Grubhofer, N. : In Radola, B. J. (ed.) *Elektrophorese Forum '80*,Bode, Munchen, p. 81 (1980).
7. Maurer, H. R. : In Maurer, H. R. *Disc Elelectrophoresis*, de Gruyter, Berlin, P. 72 (1971).
8. Maurer, H. R. and Allen, R. C. : *Z. klin. Chem. klin. Biochem. 10*, 220 (1972).
9. McManus, J. F. A. and Hoch-Ligeti, C. : *Lab. Invest. 1*, 19 (1952).
10. Allen, R. C., Spicer, S. S. and Zehr, D. : *J. Histochem. Cytochem. 24*, 908 (1976).
11. Kornfeld, R. and Ferris, C. : *J. Biol. Chem. 250*, 2614 (1975).
12. Caldwell, R. C. and Pigman, W. : *Arch. Biochem. Biophys. 110*, 91 (1965).
13. Rennert, O. M. : *Nature 213*, 1133 (1967).
14. Prescott, A. B. : *Chem. News* London *42*, 31 (1880).
15. Nauta, W. J. H. and Gygax, P. A. : *Stain Technol. 26*, 5 (1951).
16. Wray, W., Boulikas, T., Wray, V. P. and Hancock, R. : *Anal. Biochem. 118*, 197 (1981).
17. Merril, C. R., Goldman, D. and Van Keuren, M. L. : *Electrophoresis 3*, 17 (1982).
18. Sammons, D. W., Adams, L. D., Vidmar, T. J., Jones, D. H., Hatfield, C. A., Chuba, P. J. and Crooks, S. W. : In Celis, J. E. and Bravo, R. (eds.), *Two Dimensional Electrophoresis of Proteins: Methods and Applications*, Academic Press, N. Y. (in press).

18\. Marshall, T. and Latner, A. L. : *Electrophoresis 2*, 228 (1981).

19\. Switzer, R. C. III, Merril, C. R. and Shifrin, S. : *Anal. Biochem. 98*, 231 (1979).

21\. Allen, R. C. : *Electrophoresis 1*, 32 (1980).

22\. Guevera, J. Jr., Johnston, D. A., Ramagli, L. S., Martin, B. A., Capetillo, S. and Rodriques, L. W. : *Electrophoresis 3*, 197 (1982).

23\. Allen, R. C. and Arnaud, P. : *Electrophoresis 4*, 205 (1983).

24. Ansorge, W. : In Stathakos, D. (ed.), *Electrophoresis '82*, de Gruyter, Berlin, p. 235 (1983).

25. Guillemette, J. G. and Lewis, P. N. : *Electrophoresis 4*, 92 (1983).

26. Goldman, D. and Merril, C. R. : *Electrophoresis 3*, 24 (1982).

27. Adams, L. D. and Sammons, D. W. : in Allen, R. C. and Arnaud, P. *Electrophoresis '81*, de Gruyter, Berlin, p. 155 (1981).

28. Allen, R. C. : *Trends in Anal. Chem. 2*, 206 (1983).

29. Willoughby E. W. and Lambert, A. : *Anal. Biochem. 130*, 353 (1983).

30. Budowle, B. : *Electrophoresis* In Press.

31. Poehling, H. M. and Neuhoff, V. : *Electrophoresis 1*, 90 (1980).

32. Tsai, C. M. and Frausch, C. E. : *Anal. Biochem. 119*, 115, (1981).

33. Oakley, B. R., Kirsch, D. R. and Morris, N. R. : *Anal. Biochem. 105*, 361 (1980).

34. Merril, C. R., Goldman, D., Sedman, S. A. and Ebert, M. H. : *Science 211*, 1437 (1981).

35. Sammons, D. W., Adams, L. D. and Nishizawa, E. E. : *Electrophoresis 2*, 135 (1981).

36. Allen, R. C. : Unpublished observations.

37. Halt, S. J. : *Nature 170*, 271 (1952).

38. Barnett, R. and Seligman, A. : *Science 114*, 579 (1951).

39. Halt, S. J. and Withers, R. F. J. : *Nature 170*, 1012 (1952).

40. Guibault, G. G. and Kramer, D. N. : *Analyst. Chem. 37*, 120 (1965).

41. Huggins, C. and Talalay, P. : *J. Biol. Chem. 159*, 339 (1945).

42. Hopkinson, D. A., Spencer, N. and Harris, H. : *Am. J. Human Genet. 16*, 141 (1964).

43. Brewer, G. J. and Singh, Ch. F. : *An Introduction to Isozyme Techniques*, Academic Press, New York (1970).

44. Harris, H. and Hopkinson, D. A. : *Handbook of Enzyme Electrophoresis in Human Genetics*, North Holland, Amsterdam (1976).

45. Burstone, M. S. : *J. Nat. Cancer Inst. 18*, 167 (1957).

46. Hunter, R. L. and Burstone, M. S. : *J. Histochem. Cytochem. 6*, 396 (1958).

47. Allen, R. C., Popp, R. A., Moore, D. J. : *J. Histochem. Cytochem. 13*, 249 (1965).

48. Harris, H. and Robson, E. B. : *Biochem. Biophys. Acta 73*, 649 (1963).

49. Friedman, O. M. and Seligman, A. M. : *J. Am. Chem. Soc. 72*, 624 (1950).

50. Gomori, G. : *J. Lab. Clin. Med. 37*, 526 (1951).

51. Allen, R. C., Gale, G. G. and Simmons, M. A. : *Electrophoresis 2*, 114 (1982).

52. Allen, R. C., Sannes, P. L., Spicer, S. S. and Hong, C. C. : *J. Histochem. Cytochem. 28, 947 (1980).*

53. *Scopes, R. K. : Nature, 201,* 924 (1964).

54. Kuhn, R. and Jerchel, D. : *Ber. dtsch. Chem. Ges. 74B,* 941 (1941).

55. Burstone, M. S. : *Enzyme Histochemistry,* Academic Press, New York (1962).

56. Latner, A. L. and Skillen, A. W. : *Proc. Assoc. Clin. Biochem. 2,* 3 (1962).

57. Latner, A. L. and Skillen, A. W. : *Isoenzymes in Biology and Medicine,* Academic Press (1968).

58. Rothe, G. M. and Purkhandaba, H. : *Electrophoresis 3,* 43 (1982).

59. Grell, E. H., Jacobson, K. B. and Murphy, J. B. : *Science 149,* 80 (1965).

60. Martin, W. and Neibuhr, R. : *Blut 26,* 151 (1973).

61. Kinzkofer, A. and Radola, B. J. : *Electrophoresis 4,* 408 (1983).

62. Narayanan, K. R. and Raj, A. S. : in Radola, B. J. and Graesslin, D. (eds.), *Electrophoresis and Isotachophoresis* de Gruyter, Berlin, P. 221 (1977).

63. Pretch, W., Charles, D. J. and Narayanan, K. R. : *Electrophoresis 3,* 142 (1982).

64. Kinzkofer, A. and Radola, B. J. : in Radola, B. J. (ed.), *Elektrophorese Forum '82* Bode, München, p. 66 (1982).

65. Alper, C. A. and Johnson, A. M. : *Vox Sang. 17,* 445 (1969).

66. Ritchie, R. F. and Smith, R. : *Clin. Chem. 22,* 1982 (1976).

67. Sun, T., Lien, Y. Y. and Degnan, T. : Amer. J. Clin . Pathol. 72, 5 (1979).

68. Arnaud, P., Chapuis-Cellier, C., Wilson, G. B., Koistenin, J., Allen, R. C. and Fudenberg, H. H. : in Radola, B. J. and Graesslin, D. (eds.), *Electrophoresis and Isotachophoresis*de Gruyter, Berlin, p. 265 (1977).

69. Nakane, P. and Pierce, J. : *J. Cell Biol. 33,* 307 (1967).

70. Saravis, C. A., Cunningham, C. G., Marasco, P. V., Cook, R. B. and Zamcheck, N. : in Radola, B. J. (ed.), *Electrophoresis '79* de Gruyter, Berlin, p. 117 (1980).

71. Saravis, C. A., Cantarow, W., Marasco, P. V., Burke, B. and Zamcheck, N. : *Electrophoresis 1,* 191 (1980).

72. Saravis, C. A., Cook, R. B., Polvino, W. S. and Sampson, C. E. : *Electrophoresis 4,* 367 (1983).

73. Polvino, W. J., Saravis, C. A., Sampson, C. E. and Cook, R. B. : *Electrophoresis 4,* 368 (1983).

74. Southern, E. M. : *J. Mol. Biol. 98,* 503 (1975).

75. Merril, C. R. Personal Communication

76. Tollaksen, S. L., Anderson, N. L. and Anderson, N. G. : *ANL-BIM-81-1,*Argonne National Laboratory (1981).

77. Laskey, R. A. : in Grossman, L and Moldave, K. (eds.) *Methods in Enzymology 65*, Academic Press, New York, p. 363 (1980).

78. Hames, B. D. : in Hames, B. D. and Rickwood, D. *Gel Electrophoresis of Proteins: a practical approach*, IRL Press Limited, Oxford and New York, p. 52 (1981).

79. Chamberlin, J. P. : *Anal. Biochem. 98*, 132 (1979).

80. McConkey, E. H. : *Anal. Biochem. 96*, 39 (1979).

81. McConkey, E. H. : *Electrophoresis 5*, In Press.

82. Crowle, A. J. and Jewell Cline, L. : *J. Immunol. Methods 17*, 379 (1977).

83. Allen, R. C. : in Allen, R. C. and Maurer, H. R. (eds.), *Electrophoresis and Isoelectric Focusing in Polyacrylamide Gel* de Gruyter, Berlin, p. 287 (1974).

84. Allen, R. C. : *J. Chromatogr. 146*, 1 (1978).

85. Yakin, H. M., Kronberg, H., Zimmer, H-G. and Neuhoff, V. : *Electrophoresis 3*, 244 (1982).

86. Zeineh, R. A. and Abdul-Karim, K. : *Electrophoresis '83 - Boston* (1983).

87. Jansson, P. A. : *Anal. Chem. 56*, in press (1984).

AGIF	agarose isoelectric focusing
AP	ammonium persulfate
BAC	bisacrylcystamine
Bis	N,N'-methylenebisacrylamide
C	degree of cross linking as a per cent of T
CHAPS	zwitterionic detergent
Con-A	concanavalin A
Cp	ceruloplasmin
CRP	C reactive protein
CSF	cerebral spinal fluid
CZE	continuous zone electrophoresis
DATD	N,N_1-diallyltartardiamide
DCIP	dichloroindophenol
DHEBA	N,N'- (1,2-dihydroxyethylene) bisacrylamide
Disc	discontinuous electrophoresis, normally meaning the Ornstein - Davis system
DMSO	dimethylsulfoxide
DOC	desoxycholate
DTT	dithiothreitol
EDA	ethylene diacrylate
EDTA	ethylenediaminetetraacetic acid, disodium salt
f	discharge rate
Gc	group specific protein
GPT	glutamic-pyruvate-transaminase
Hb	hemoglobin
HDL	high density lipoprotein
HLA	human lymphocyte antigens
Hp	haptoglobin
HRP	horseradish peroxidase
IEF	isoelectric focusing
K_d	dissociation constant
KP	potassium persulfate

K_R	retardation coefficient
LDH	lactic acid dehydrogenase
LDL	low density lipoprotein
βME	β-mercaptoethanol
MTT	(3-(4,5-dimetylthiazole) -2,5-diphenyltetrazolium chloride
MW	molecular weight
MZE	multizonal electrophoresis
NBT	nitro blue tetrazolium chloride
NEPHGE	non-equilibrium pH gradient electrophoresis
PAGE	polyacrylamide gel electrophoresis
PAGIF	polyacrylamide gel isoelectric focusing
PAP	peroxidase antiperoxidase
PAS	periodic acid Schiff stain
PEGDA	polyethyleneglycol diacrylate
PGM	phospho-gluco-mutase
Pi	proteinase inhibitor
pI	isoelectric point
PKU	phenylketonuria
PMS	N-methyl-phenizine-methylsulfate
PPO	2,5-diphenyloxazole
R_f	relative mobility
RN	riboflavin
R_s	resolving power IEF
SCAM	synthetic carrier ampholyte
SDS	sodium dodecylsufate
T	per cent monomer and crosslinking agent
TBS	tris buffered saline
TCA	trichloroacetic acid
TEMED	N,N,N^1,N^1 -tetramethylethylenediamine
VLDL	very low density lipoprotein
Y_0	y intercept

5.14. INDEX

E

F

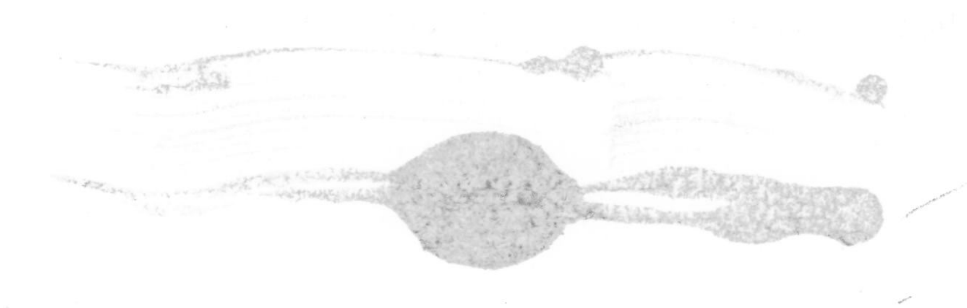